EXPERIMENTAL TECHNIQUES AND DESIGN IN COMPOSITE MATERIALS 4

PROCEEDINGS OF THE 4th SEMINAR ON EXPERIMENTAL TECHNIQUES AND DESIGN IN COMPOSITE MATERIALS – SHEFFIELD, U.K., 1-2 SEPTEMBER 1998

Experimental Techniques and Design in Composite Materials 4

Editor
Michael S. Found
University of Sheffield, United Kingdom

A.A. BALKEMA PUBLISHERS   LISSE / ABINGDON / EXTON (PA) / TOKYO

Cover figure:
N. Petrone and M. Quaresimin

Published by: A.A. Balkema, a member of Swets & Zeitlinger Publishers
                 www.balkema.nl and www.szp.swets.nl

ISBN 90 5809 370 0

Printed in the Netherlands

Experimental Techniques and Design In Composite Materials 4, Found (Ed.)
© *2002 Swets & Zeitlinger, Lisse, ISBN 90 5809 370 0*

Contents

Section 4: Impact

Section 5: Modelling

Experimental Techniques and Design In Composite Materials 4, Found (Ed.)
© 2002 Swets & Zeitlinger, Lisse, ISBN 90 5809 370 0

Preface

This volume contains the revised versions of papers presented at the Fourth Seminar on Experimental Techniques and Design in Composite Materials held in Sheffield on 1-2 September 1998 and hosted by the Structural Integrity Research Institute (SIRIUS) and the Advanced Railway Research Centre (ARRC) of the University of Sheffield. These seminars are held biennially with the venue alternating between Cagliari and Sheffield with the aim of bringing together engineers, materials scientists, designers and end users in order to exchange knowledge and encourage wider use of composite components and structures.

The papers have been divided into five sections: Fatigue, Test Methods, Design, Impact and Modelling. They identify the need for continuing careful experimentation in order to support modelling and simulation methods to enhance the use of composites with improved designs through applications in more industries. There are some significant developments in modelling and simulation covering a range of problems from small scale modelling of the stress transfer of single fibres to large scale structural analysis. It is interesting to note that the difficulties, assumptions and limitations are often similar, whatever the scale of the problem. Some interesting structural designs are presented for transport applications ranging from bumpers and stabiliser bars for the automotive industry to a kayak for Olympic racing and to bodyshells for rail vehicles. Whilst there has been significant progress over recent years towards the understanding of the behaviour of composites there is still much to be achieved if these materials are to evolve into more general use. However, it is the use of high-performance composites that are driving many of these advances, together with developments in processing, which are likely to lead to improvements in performance, structural integrity and reliability. In terms of future developments the following appear to be of importance in assisting the further advance for designs in composite materials:

- modelling and simulation techniques at both micro and macro scales
- advances in experimental techniques
- validation of analytical solutions for components and structures in service
- modelling the progressive development of damage in its various forms

The authors of the papers have provided a significant contribution to a better understanding of some of the issues that were addressed in the seminar and we hope that they will inspire others to use these evolving technologies to produce innovative designs in composite materials.

The success of the seminar and the production of the proceedings would not have been possible without the hard work and support of a number of individuals. In particular, grateful thanks are expressed to members of the Scientific Committee for refereeing of the papers and for the valuable help provided by Mrs N A Parkes.

Michael S Found

Section 1: *Fatigue*

Experimental Techniques and Design In Composite Materials 4, Found (Ed.)
© 2002 Swets & Zeitlinger, Lisse, ISBN 90 5809 370 0

Guidelines for fatigue design of fibre-reinforced metal matrix composites

E.R. de los Rios, C.A. Rodopoulos & J.R. Yates
University of Sheffield - Department of Mechanical Engineering
Mappin Street, Sheffield S1 3JD U.K.

ABSTRACT: Composite Fracture Mechanics principles are used to model fatigue crack growth in MMCs. In its simplest form composite modelling considers the material as a composite of two components, matrix and fibre, with a crack system divided into three zones: the crack, the plastic zone and the fibre constraint zone. The solution of the model equations allows for the calculation of the stresses sustained by the crack wake, plastic zone, barrier zone and elastic enclave. It also leads to the calculation of the crack open displacement (COD) over the entire crack length and of the crack tip open displacement (CTOD). Crack growth rate is calculated through a Paris type relationship in terms of CTOD, i.e. da / dN = C × CTOD^m. Conditions for crack arrest and instability are established and used to derive fundamental tools for damage tolerance design.

1 INTRODUCTION

Composite Fracture Mechanics is a spin-off of Microstructural Fracture Mechanics which deals with fracture problems where the microstructure of the material plays a significant part. One significant area is the study of short fatigue cracks in which important developments have taken place in the past few years.

These developments have demonstrated the considerable effect that a materials microstructure has on the early stages of crack growth (Miller 1987), and have introduced the concept of microstructural barriers to crack propagation and their connection with the fatigue limit (de los Rios et al. 1984 & 1985). Work in this area has also demonstrated the importance of crack tip plastic deformation in the crack propagation rate (Chang et al. 1979, Weertman 1966).

The understanding of these two aspects i.e. barrier strength and crack tip plasticity, and their influence on crack growth, has led to the evolution of short crack models from an empirical beginning (Hobson et al. 1986) to a more fundamental approach (Navarro & de los Rios 1988a). It is now possible to incorporate the mechanical driving force of crack growth, represented by the applied stress and crack length, and the material resistance in a explicit form within the crack system equations (Navarro & de los Rios 1992, de los Rios et al. 1994).

The material resistance is of two kinds, one is intrinsic and considers the material parameters controlling crack tip plastic flow such as those which determine the strength of the barriers such as the differential strength of constituent phases and the transfer of plasticity across barriers. The other resistance encountered as the crack grows concerns the interaction between the crack faces such as crack locking or bridging and is a function of both the friction forces on the faces, which is dominated by crack surface topography, and Mode I crack tip plasticity, which controls crack face separation.

2 MICRO-MECHANICAL MODELLING OF CRACK GROWTH

The model crack system is composed of three zones, the crack, the plastic zone and the fibre constraint zone. These three zones are simulated by distribution of dislocations as modelled by Bilby, Cottrel and Swinden (Bilby et al. 1963). The distribution is of screw dislocations for Mode III loading, edge dislocations with the Burger's Vector parallel to the crack plane for Mode II and edge dislocations with the vector perpendicular to the crack plane for Mode I loading. Navarro and de los Rios have solved the equilibrium equation for a multizone crack system (Navarro & de los Rios 1988b) and applied it to the three zone case (Navarro & de los Rios 1988c).

When the model is applied to fibre-reinforced MMCs (de los Rios et al. 1996a & 1996b) it considers a fatigue crack in a composite of interfibre spacing D and fibre diameter d with the crack tip plastic zone being blocked at the fibre zone (see Fig. 1). The solution of the equilibrium equation provides the means of calculating the distribution of stresses over the three zones; which are σ_1 at the crack, σ_2 at the plastic zone and σ_3 at the fibre constraint zone; (see Fig. 2) and the expression for the crack open displacement, COD. The stress σ_1 is the bridging stress, σ_2 is the yield stress of the plastic zone and σ_3 is calculated as:

$$\sigma_3 = \frac{1}{\cos^{-1} n_2}\left[(\sigma_2 - \sigma_1)\sin^{-1} n_1 - \sigma_2 \sin^{-1} n_2 + \frac{\pi}{2}\sigma\right] \tag{1}$$

and the crack open displacement as:

$$
\begin{aligned}
\text{COD} = \frac{bc}{\pi^2 A}\Bigg\{ & (\sigma_2 - \sigma_1)\left[(\zeta_b - n_1)\cosh^{-1}\left(\left|\frac{1 - n_1\zeta_b}{n_1 - \zeta_b}\right|\right) - (\zeta_b + n_1)\cosh^{-1}\left(\left|\frac{1 + n_1\zeta_b}{n_1 + \zeta_b}\right|\right)\right] \\
& -(\sigma_2 - \sigma_1)\left[(\zeta_a - n_1)\cosh^{-1}\left(\left|\frac{1 - n_1\zeta_a}{n_1 - \zeta_a}\right|\right) - (\zeta_a + n_1)\cosh^{-1}\left(\left|\frac{1 + n_1\zeta_a}{n_1 + \zeta_a}\right|\right)\right] \\
& +(\sigma_3 - \sigma_2)\left[(\zeta_b - n_2)\cosh^{-1}\left(\left|\frac{1 - n_2\zeta_b}{n_2 - \zeta_b}\right|\right) - (\zeta_b + n_2)\cosh^{-1}\left(\left|\frac{1 + n_2\zeta_b}{n_2 + \zeta_b}\right|\right)\right] \\
& -(\sigma_3 - \sigma_2)\left[(\zeta_a - n_2)\cosh^{-1}\left(\left|\frac{1 - n_2\zeta_a}{n_2 - \zeta_a}\right|\right) - (\zeta_a + n_2)\cosh^{-1}\left(\left|\frac{1 + n_2\zeta_a}{n_2 + \zeta_a}\right|\right)\right]\Bigg\}
\end{aligned}
\tag{2}
$$

where σ is the applied stress; b the Burgers vector, A = $Gb/2\pi$ for screw dislocations, $A=Gb/2\pi$ (1-υ) for edge dislocations, υ Poisson's ratio and G the shear modulus. The other variables are defined in Fig. 1.

If crack growth rate is considered to be a function of crack tip open displacement, CTOD, through a Paris type relationship (see equation 3), then equation (2) determines da/dN when $\zeta_a = n_1$, and $\zeta_b = 1$.

At critical crack lengths in each interfibre spacing, σ_3 attains the value required to overcome the next fibre constraint, plasticity then spreads to the next fibre and the crack is able to propagate through the fibre zone. Alternatively, if the applied stress is below a critical value (endurance limit), then σ_3 will not attain the value required to overcome the fibre constraint and the crack arrests. Because crack tip plastic displacement is pro-

portional to the extent of plasticity at the crack tip, it shows an oscillating variation within each fibre spacing. When a barrier is overcome, its value is high, but as the crack grows towards the next fibre, its value decreases to a minimum at the critical crack length (see Fig. (3)).

2.1 Crack propagation rate

Crack propagation rate is calculated by a Paris type relationship using the crack tip open displacement as the driving force:

$$\frac{da}{dN} = C \times CTOD^m \tag{3}$$

The scale parameters C and m are obtained by a numer-

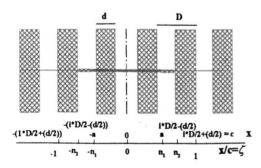

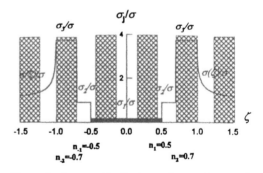

Figure 1. Schematic representation of the crack system. The crack length is 2a, the fibre diameter is d and the fibre spacing is D. On the positive coordinates side the crack tip is at a, the plastic zone extends to the next fibre at i × D/2 - (d/2) and the constrained zone extends to i × D/2 + (d/2), i = 1, 3, 5 ...

Figure 2. Stress distribution along the crack system. The stress at the crack itself is σ_1 (bridging stress), the stress at the plastic zone is the flow stress σ_2, the stress at the constrained zone is σ_3 and $\sigma(\zeta)$ is the elastic stress distribution.

4

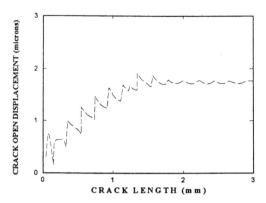

Figure 3. Variation of crack tip open displacement for a propagating crack, showing the effect of fibre constraint at short crack lengths (less than 2 mm) and the negligible effect of fibre constraint at longer crack lengths.

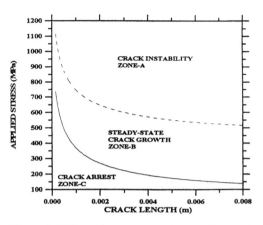

Figure 4. Fatigue damage map showing the crack arrest, crack propagation and crack instability regions (σ_{3d} = 1173 MPa and matrix yield stress σ_y = 556 MPa).

ical method relating the integration of equation (3) between initial and final crack length and the S-N curve of the matrix or by using long crack propagation data and CTOD values corresponding to cracks longer than 6-8 interfibre spacings, which would be beyond the fibre constraint sensitive region of crack growth.

The initiation of the crack is a process believed to be very rapid since there are intrinsic stress concentrations in these type of materials. As the crack grows, the value of the stress at the barrier σ_3 increases such that one of two possibilities develops. If σ_3 does not attain the value to overcome the fibre constraint, the crack tip will reach the grain boundary and will arrest. Conversely, at a critical crack length, σ_3 attains the required value, the fibre constraint is overcome, crack tip plasticity extends into the next fibre and the crack will be able to propagate to failure unless the applied stress is decreased or there is a build up of crack bridging (increases in σ_1). The applied stress may decrease as in variable amplitude loading while crack bridging may increase due to the extra number of fibres left behind in the crack wake. However, if the loading is of constant amplitude and bridging fibres fracture readily, the crack would propagate in a stable manner up to the instability stage.

2.2 Fatigue damage map

For the application of this micro-mechanical model to engineering practice it is important to be able to establish the bounds of the different damage zones. In general there are three main damage regions in fatigue: the crack arrest region, the crack propagation region and the crack instability region, see Fig. 4.

2.3 Crack arrest

In this region even if a fatigue crack is initiated, it will come to arrest because the applied stress is lower than the endurance limit of the material. The conditions for crack arrest in monolithic materials have been determined experimentally and presented in plot form by Kitagawa and Takahashi (Kitagawa & Takahashi 1976). In fibre reinforced MMCs, the situation is more complicated because the endurance limit for a given defect size, depends on the number of fibres bridging the crack. This number will be a function of the probability of fibre failure for any fibre in the crack wake.

Within the concepts of the micro-mechanical model described above, the crack arrests when the stress σ_3 does not reach the level required to extend plasticity across the fibre constraint zone before the crack tip reaches the fibre. The condition for arrest can be obtained from Equation (1) by considering that when the tip of the crack reaches the fibre before plasticity spreads to the next fibre spacing $n_1 = n_2 \cong 1$. Thus, the stress for crack arrest is:

$$\frac{2}{\pi}\sigma_3^i \cos^{-1} n_2^i + \sigma_1 = \sigma_{arrest} \qquad (4)$$

where " i " is the number of half inter-fibre spacings within the crack system. By using the approximation

$$cos^{-1} n_2^i \approx \left(2\left(1 - n_2^i\right)\right)^{1/2} \approx \sqrt{2\sqrt{d/(a+d)}} \qquad \text{and}$$

considering that the maximum stress that causes arrest is obtained where $\sigma_3^i = \sigma_{3d}$ where σ_{3d} is the stress required at the fibre zone to debond a particular fibre length L.

Then the maximum allowed applied stress which would still lead to the arrest of a crack of length a is:

$$\sigma_{arr} = \frac{2\sqrt{2}}{\pi}\sigma_{3d}\sqrt{\frac{d}{a+d}} + \sigma_1 \qquad (5)$$

5

The stress σ_{3d} can be written as,

$$\sigma_{3d} = \frac{4\,\tau\,L\,E_c}{d\,E_f} \qquad (6)$$

where E_c and E_f are the elastic moduli of the composite and the fibre respectively, L is the embedded fibre length, d is the fibre diameter and τ is the interfacial shear strength.

Published results in the literature indicate values for interfacial shear strength of SCS6/Ti-15-3 MMC in the range of 124 to 148MPa (Yang et al. 1991 & Mackin et al. 1992).

The effect of a bridging stress (σ_1) is that of increasing the stress necessary for arrest i.e. the arrest curve moves upwards.

2.4 Crack Propagation

For applied stress higher than the endurance limit, the crack will propagate in a steady manner up to failure (instability). The integration of equation (3) would determine the number of cycles required for crack to grow between two given limits. The designer could use this procedure to ascertain that the component would not fail within the design life. The maintenance engineer likewise, could perform calculations based on the integration of equation (3) to establish inspection intervals.

As shown in Fig. (3), fibre constraints effects as characterised by oscillations in the crack growth rate curve, are prominent at short crack length, while beyond this region the oscillations are of negligible intensity and therefore, fibre constraint does not then offers any significant resistance to crack propagation. Most of the resistance is then borne by the bridging stress. The increase in bridging stress as the crack gets longer (more fibres bridging the crack), compensate the raising crack driving force (longer crack), with the net result of attaining a constant crack propagation rate.

The theoretical expression that establishes the transition from fibre constraint dominated crack growth, to a more continuous type growth, was derived in de los Rios et al. 1998 and predicts that this occurs when the size of the plastic zone is larger than one interfibre spacing. The assumption is that the crack tip stresses are of sufficient magnitude to debond the next two fibre-rows, thus considering that the critical state for a high degree of constraint effect degradation could be formulated as (iD/2-d/2)-a ≥ 2D, for $\sigma_3 = \sigma_{3d}$, Eq (1) yields,

$$\sigma_{cons} = \frac{2\sqrt{2}}{\pi}\left(\sigma_{3d}\sqrt{\left(1 - n_2^{\,2}\right)} - \sigma_2 n_1 + \sigma_2 n_2\right) \qquad (7)$$

where $n_1 = \dfrac{a}{a + 2D}$ and $n_2 = \dfrac{a + 2D - d}{a + 2D}$. In Eq (7)

the closure stress σ_1 is considered negligible (wide spread failure of the bridging fibres).

Basically this constraint effect degradation curve could be used as a representation of the effectiveness of the fibres (volume fraction, interfacial shear strength and fibre diameter) on providing fatigue strength in MMCs, in a manner similar to the effect of the grain boundary properties on the transition from microstructural sensitive to microstructural insensitive fatigue of monolithic materials.

2.5 Instability

The solution of the stress equilibrium equation (Eq. (1)), gives the position of the crack tip in relation to the fibre when the fibre constraint is overcome as:

$$n_1^c = cos\left(\frac{\pi}{2}\,\frac{\sigma - \sigma_{Li}}{\sigma_2 - \sigma_1}\right) \qquad (8)$$

This expression shows that for stable crack growth the material resistance ($\sigma_2 - \sigma_1$) should remain larger than the crack driving force ($\sigma - \sigma_{Li}$) where σ_{Li} is the threshold stress for crack propagation (equation 5). However, a state will be reached at a given crack length when the material resistance can no longer compensate for the increase in crack driving force, thus

$$\frac{\sigma - \sigma_{Li}}{\sigma_2 - \sigma_1} \to 1 \qquad (9)$$

and then from equations (7) and (8).

$$\sigma_{ins} = \frac{2\sqrt{2}}{\pi}\,\sigma_{3d}\sqrt{\frac{d}{a + d + a_{in}}} + \sigma_2 \qquad (10)$$

Equation (10) is plotted in Fig. (4) as the upper limit for the fatigue crack propagation region. The first term of the left hand side represents the stress required to overcome the fibre constraint. It shows that for long cracks this term can be neglected, which confirms that for long cracks the effect of fibre constraint is negligible and instability is achieved when $\sigma_{ins} = \sigma_2$, which is Bilby's (Bilby et al. 1963) condition for general yielding from a notch.

3 DISCUSSION AND CONCLUSIONS

The growth of fatigue cracks in fibre-reinforced MMCs, initiated at surfaces without obvious defects, requires comparatively higher stresses than that required by long cracks growing in Fracture Mechanics specimens. This means that in most instances the condition of small scale yielding for linear elastic fracture mechanics (LEFM) will be violated. This condition is violated because the relatively large cyclic plasticity at small crack tips alters significantly the strength of stress field ahead of the fatigue crack and thus Elastic Plastic Fracture Mechanics (EPFM) solutions are needed.

In this work the solution developed by Bilby (Bilby et al. 1963) to obtain the conditions for plastic yielding from cracks or notches is used as the basis for developing EPFM solutions of fatigue crack growth. The effect of fibre reinforcements is subsequently incorporated into these solutions to set up the principles on which fatigue modelling of fibre reinforced MMCs is based.

The presence of fibres in composites may affect the fatigue resistance of the material in various ways. Fibres are shown to constrain crack tip plasticity and also prevent the crack of opening fully by creating friction forces at the crack faces (bridging stresses). These effects are incorporated into the micro-mechanics model through the development of a crack closure stress σ_1 and a constraint stress σ_3.

Grain size or any other microstructural constituent affecting the flow characteristics of the matrix are included in the model through the σ_2 term. For example matrix heat treatments and grain refinement, or the cyclic strain hardening or softening produced at the crack tip plastic zone, could readily be included in crack growth rate or fatigue life calculations to grade composites according to their matrix characteristics. Grain boundaries, second phases and reinforcements (i.e. fibres) also affect the fatigue process by their effect in constraining crack tip plasticity. Crack tip plasticity displacements are prevented at the matrix reinforcements (i.e. fibres) which generate a stress σ_3 at that location. The crack will be able to propagate across the fibre zone once the stress σ_3 achieves a critical value which is related to the fibre-matrix bonding strength. Methods developed by the materials engineer to strengthen materials, involving fibre-matrix interfaces, could easily be assessed in terms of their fatigue resistance. Conversely, composite fracture mechanics could be used to issue advice and guidelines for the development of a new generation of crack resistance composites.

Finally, the methodology presented here could assist the design and maintenance engineer to perform damage and lifetime calculations. Basic fatigue data can be used to predict the behaviour of the material when a new manufacturing route is planned. A fatigue damage map could be constructed for the determination of safe loads showing the particular design approach which could best be used. If a safe-life methodology is used, the design loads should be bounded by the arrest curve. However, if a damage tolerance approach is used, loads in the steady crack growth region are possible and calculations must be performed to ascertain the safety of the components and the inspections intervals.

In summary, composite fracture mechanics models incorporating material and mechanics variables can be used to develop guidelines for design and safety engineering, and to predict the effect of changes to manufacturing processes on the fatigue properties of composite components, thus reducing the requirement of extensive testing programmes.

REFERENCES

Bilby B.A., Cottrell A.H. and Swinden K.H., The Spread of Plastic Yield from a Notch, *Proc. R. Soc. London, A272,* (1963), 304

Chang R., Morris W.L. and Buck O., Fatigue Crack Nucleation at Intermetallic Particles in Alloys - A Dislocation Pile-up Model, *Scripta Metall.* 13, 1979, 191-

de los Rios E.R., Tang Z. and Miller K.J., Short Crack Fatigue Behaviour in a Medium Carbon Steel, *Fatigue Engng. Mater. Struct.*, 7, 1984, 97-108.

de los Rios E.R., Mohamed A.J. and Miller K.J., A Micro-mechanics Analysis for Short Fatigue Crack Growth, *Fatigue Fract. Engng. Mater. Struct.*, 8, 1985, 49-64.

de los Rios E.R., Xin X.J. and Navarro A., Modelling Microstructural Sensitive Fatigue Short Crack Growth, *Proc. R. Soc. Lond.*, A447, (1994), 111-134.

de los Rios E.R., Rodopoulus C.A. and Yates J.R., A model to predict the fatigue life of fibre reinforced titanium matrix composites under constant amplitude loading. *Fatigue Fract. Engng. Mater. Struct.* 19, 539-550, 1996.

de los Rios E.R., Rodopoulus C.A. and Yates J.R., Modelling the conditions for fatigue failure in metal matrix composites. *Fatigue Fract. Engng. Mater. Struct.* 19, 1093-1105, 1996.

de los Rios E.R., Rodopoulus C.A. and Yates J.R., The fatigue behaviour of metal matrix composites under single overloads. *Fatigue Fract. Engng. Mater. Struct.* 21, 1503-1511, 1998.

Hobson P.D., Brown M.W. and de los Rios E.R., Two Phases of Short Crack Growth in a Medium Carbon Steel. *The Behaviour of Short Fatigue Cracks*, (Mech. Engng. Publs. London, 1986), 441-459.

Kitagawa H. and Takahashi S., Applicability of fracture mechanics to very small cracks or cracks in the early stage. *2nd Int. Conf. on the Mechanical Behaviour of Materials (ICM2)*, (Am. Soc. Metals, Metals Park, Ohio, 1976), 627-631.

Mackin T.J., Warren P.D. and Evans A.G. (1992) Effects of fibre roughness on interface sliding in composites, *Acta Metall. Mater.* Vol. 40, 1251-1257.

Miller K.J., The Behaviour of Short Fatigue Cracks and their Initiation - Part I: A Review of Two Recent Books, *Fatigue Fract. Engng. Struct.*, 10 (1987), 75-91. Part II: A General Summary, *Fatigue Fract. Engng. Struct.*, 10 (1987), 93-113.

Navarro A. and de los Rios E.R., Short and Long Fatigue Crack Growth. A Unified Model, *Phil. Mag. A*, 57, (1988a), 15-36.

Navarro A. and de los Rios E.R., Fatigue Crack Growth Modelling by Successive Blocking of dislocations, *Proc. R. Soc. Lond.*, A437, (1992), 375-390.

Navarro A. and de los Rios E.R., Compact Solution for a Multizone BCS Crack Model with Bounded or Unbounded end Conditions. *Phil. Mag. A*, 57, (1988b), 43-50.

Navarro A. and de los Rios E.R., An alternative Model of the Blocking of Dislocations at Grain Boundaries. *Phil. Mag. A*, 57, (1988c), 37-42.

Weertman J., Rate of Growth of Fatigue Cracks Calculated from the Theory of Infinitesimal Dislocation Distributed on a Plane, *Int. J. Fracture Mech.*, 2, 1966, 460-

Yang J.M., Jeng S.M. and Yang C.J. (1991) Fracture mechanisms of fibre-reinforced titanium alloy matrix composites-Part I: Interfacial behaviour, *Mat Sci. Engng.* A138, 155-167.

Experimental Techniques and Design In Composite Materials 4, Found (Ed.)
© 2002 Swets & Zeitlinger, Lisse, ISBN 90 5809 370 0

Fatigue Damage in a Glass Fibre Reinforced Polypropylene Composite

P.N.B. Reis
Department of Electromechanical Engineering, UBI, Covilhã, Portugal

J.A.M. Ferreira
Department of Mechanical Engineering, FCTUC, University of Coimbra, Coimbra, Portugal

J.D.M. Costa
Department of Mechanical Engineering, FCTUC, University of Coimbra, Coimbra, Portugal

M.O.W. Richardson
IPTME, Loughborough University, Loughborough LE11 3TU, UK

ABSTRACT: Thermoplastic resins with high average molecular weight and high crystallinity exhibit relatively high melting temperature and mechanical properties. Advanced fibre reinforced thermoplastic composites fabrics, which consist of thermoplastic filaments interwoven with reinforced E-glass fibres have been developed recently which enable thermoplastic composites to be produced and are today good alternatives to thermosetting resin systems. The most attractive features offered by thermoplastic composites are potential low cost manufacturing, high fracture toughness, good damage tolerance and impact resistance, good resistance to microcracking, simple quality control and the possibility of recycling the raw materials.

The paper concerns fatigue studies of polypropylene/glass fibre thermoplastic composites produced from a bi-directional woven cloth of co-mingled E glass fibres and polypropylene fibres (the latter becoming the matrix after the application of heat and pressure). This composite was manufactured with a fibre volume fraction V_f of 0.338. The effect of layer design and load conditions on fatigue performance were investigated. The S-N curves, the rise in the temperature of the specimens during the tests, and the loss of stiffness, were obtained and are discussed. The loss of stiffness (E) was used as a damage parameter and related to rise in temperature and stress release observed in the material during the first period of fatigue life. The results show that the damage parameter (E) present a nearly linear relationship with the rise in temperature.

1 INTRODUCTION

Thermoplastic resins with high average molecular weight and high crystallinity exhibit relatively high melting temperatures and mechanical properties. Advanced fibre reinforced thermoplastic composite fabrics, which consist of thermoplastic filaments interwoven with reinforcing E-glass fibres have been developed recently which enable thermoplastic composites to be produced and offer good alternatives to thermosetting resin systems. The most attractive features offered by thermoplastic composites are the potential of low cost manufacturing, high fracture toughness, good damage tolerance and impact resistance, good resistance to microcracking, simple quality control and the possibility to recycle the raw materials. A variety of thermoplastic resins have been investigated as matrices, including polypropylene, nylon, polyetherimide, polyphenylene sulphide, polyetheretherketone. Hou et al. (Hou et al. 1995) reviewed the relative costs of different material forms and the manufacturing techniques required.

The most important thermoplastic applications are in the automobile industry including panels, seat frames, bumper beams, load floors, front-end structures, head lamp retainers, rocker panels, under-engine covers (Harper & Pugh 1991). Other long fibre thermoplastic applications are reported in the construction, electrical and recreational sectors. One of the most important applications of glass reinforced polypropylene is in automotive body panels by low cost thermoforming techniques.

The paper presents results associated with a glass fibre reinforced polypropylene composite where the influence of the fibre orientation and load mode on static and fatigue behaviour were investigated. The fatigue strength was obtained in terms of the number of cycles to failure versus the stress range. In previous work (Ferreira et al. 1996, Ferreira et al. 1997) other parameters such as frequency and stress concentration were reported.

The main objective of this work was to understand the damage mechanisms observed during the fatigue process. One of the fatigue damage parameter reported in the literature (Pimk & Campbell 1974, Sims & Gladman 1978, Sims & Gladman 1980) is the rise in specimen temperature, which increases with fatigue life especially close to final failure. The specific rise in temperature was measured for all the tests using a thermocouple with three probes placed in the failure region.

Other fatigue damage investigated included the residual stress, the loss of stiffness, the event number detected by ultrasonic analysis and fracture mechanics parameters. The loss of stiffness during fatigue tests has been investigated as a damage criterion by Hahn (Hahn & Kim 1976) and Joseph (Joseph & Perreux 1994) in glass/epoxy, the authors (Ferreira et al. 1996, Ferreira et al. 1997) in glass/polypropylene and Echtermeyer (Echtermeyer et al. 1995) in glass/phenolic and glass/polyester composites.

The lay-up geometry has a strong influence on fatigue strength. Curtis (Curtis 1987) and Harris (Harris et al. 1990) investigated this effect in carbon/epoxy composites. The objectives of the current study were to investigate the effect of lay-up geometry on the fatigue strength and fatigue damage of polypropylene resins reinforced by bi-directional glass fibre layers. The loss of stiffness and the rise of temperature on the surface of the specimen were used to quantify the fatigue damage.

2 MATERIAL AND EXPERIMENTAL PROCEDURES

Composite sheets were formed from multi-layers of (Vetrotex Twintex T PP) of crossed continuos E-glass which were processed in a mould under pressure (5 bar for ten minutes) after heating at 190°C. This temperature was chosen to be above the melting temperature of the polypropylene. Each sheet was made up of seven woven balanced bi-directional ply layers. The overall dimensions of the plates were 160x250x3 mm with a fibre volume fraction indicated by the manufacturer of 0.338. Quality controls of the plates include visual inspection of the colour and void content.

Three types of plate were manufactured. For one series of plates all the layers plies have one of the two fibre directions orientated with the axis of the plate. The other two plates had the following ply orientation with respect to the axis of the sheet: +45°/0°/-45°/0°/+45°/0°/-450 and +30°/-30°/+30°/0° /+30°/-30°/+300. For convenience these three types are referred as 0°, +45°/0°/-45° and +30°/-30°/0°, plates respectively.

The specimens used in the fatigue tests were prepared from these thin plates. The geometry and dimensions of the fatigue specimens are shown in Fig. 1.

The specimens used in static tests were similar but with a cross section width of 10 mm. Two types of fatigue test were performed. One series of fatigue tests were carried out in an electromechanical machine where frequency and stress ratio can be changed and the load is monitored by a load cell. The tests were performed in constant amplitude displacement mode (the load wave was sinusoidal constant amplitude). Other series of tests were carried out in a servo-hydraulic Instron machine in constant amplitude load. All the tests were performed in tension with stress ratio R = 0.025 and frequency 10 Hz at ambient temperature.

During the fatigue tests the temperature rise at three points on the surface of the specimens was measured using thermocouples and recorded in a computer. Periodically the procedures were stopped and a static loading test carried out. The stiffness modulus was derived from the linear regression of the stress-strain curves so produced.

3 RESULTS AND DISCUSSION

The tensile mechanical properties were obtained using an electromechanical Instron Universal Testing machine. The tests were carried out for the five rates of strain loading (RSL) of 0.333, 0.0333, 0.00333, 0.00033 and 0.000033 s^{-1}. For each condition, four specimens were tested. Average values were obtained for the tensile strength in the five rates of strain loading (RSL) at three different layer distributions.

One of the objectives of static tests was to study the influence of layer distribution and strain rate on the static strength. The results of the static strength σ_{UTS} are plotted in Fig. 2. As reported by Sims (Sims & Gladman 1978, Sims & Gladman 1980) for other glass fibre composites the static strength σ_{UTS} increases with the increase of RSL, although in this case this influence is very low for +45°/0°/-45° and +30°/-30°/0° laminates and more pronounced for 0° laminate (where the static strength increases about 25% when the strain rate increases from 0.0000333 to 0.333 s^{-1}). Static strength presents results very close to the +45°/0°/-45° and +30°/-30°/0° laminates and much lower than the 0° laminate (this laminate has a strength and stiffness approximately two times and 1.5 times the other laminates, respectively). The improvement of strength in the 0° laminate is caused by the failure mechanism changing.

Fig. 3 shows the failure aspect of the broken specimens for 0° and +45°/0°/-45° laminates, respectively. In the +45°/0°/45° and +30°/-30°/0° laminates the main

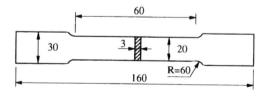

Figure 1. Fatigue test specimen (dimension in mm).

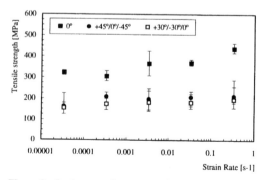

Figure 2. Static strength versus strain rate.

a)

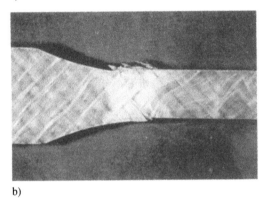

b)

Figure 3. Fracture aspect after static tests. a) 0°; b) +45°/0°/-45°.

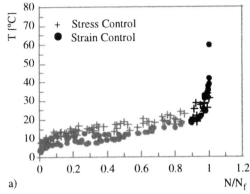

a)

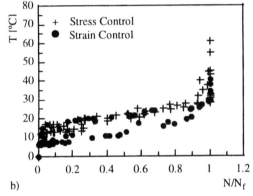

b)

Figure 4. Rises in surface temperature. a) 0°; b) +45°/0°/-45°.

failure mechanism is the delamination in the inclined fibres although in the 0° laminate the failure begins by delamination in transverse fibres but the load is transferred to the longitudinal fibres. The final failure occurs by braking of these longitudinal fibres.

The three types of laminates (0°, +45°/0°/-45° and +30°/-30°/0°) were tested in load amplitude control and displacement amplitude control During all the fatigue tests the temperature rise at three points on the surface of the specimens were measured and periodically the static stiffness modules was obtained (as described earlier). These parameters are used in this discussion to interpret and understand the damage mechanisms.

For analysis purposes the highest of the three-recorded surface temperatures was used in each case; in spite of that only a little difference between the three thermocouple values was observed until the final fatigue period. Figs. 4a) and 4b) show the increase in temperature versus the dimensionless life N/N_f (where N is the number of cycles at any given instant of the test, N_f is the number of cycles to failure) for 0° and +45°/0°/-45° laminate, respectively.

The rises of temperature associated with both load amplitude and displacement control tests were plotted. These figures show similar behaviour for both types of

laminate. The maximum temperature in both laminates occurred at failure. It can be seen that there is an initial increase of temperature followed by a stage where it is nearly stabilised and thereafter by a very small increase until close to failure when there is a sudden increase. During the second stage there is a balance between the rate of deformation energy lost by the material and the energy dissipation rate. In this period there is an inverse relationship between temperature and stiffness (Fig. 5). In the third period the rate of energy release cause by failure of the matrix and fibres is predominantly high and the temperature increases quickly. The rise in temperature is dependent on the stress range especially close to failure (that was in the range 25-75°C). However, no correlation is evident between the rise in temperature and the stress range. Similar behaviour occurs for both test loading modes (stress range or strain range control) despite a tendency to reach higher temperature values in the first case (stress range mode).

Fig. 5a) and b) plots E/E_o versus N/N_f for 0° and +45°/0°/-45° laminates, respectively, where E is the stiffness modules at any given moment of the test and E_o is the initial stiffness modulus.

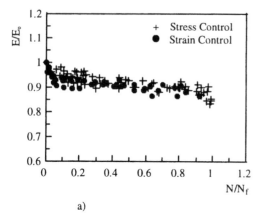

a)

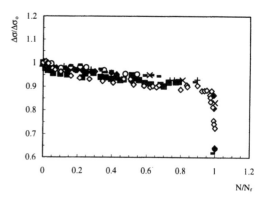

Figure 6. Δs/Δs0 against N/Nf in displacement controlled test, 0° laminate.

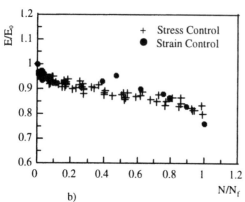

b)

Figure 5. E/E_0 plotted against the normalised number of cycles N/N_f. a) 0°; b) +45°/0°/-45°.

These figures show that in all three laminates similar behaviour occurs namely a significant drop in stiffness during the first fatigue cycles (an average drop about of 5% during the first 5% of the fatigue life). Thereafter, the stiffness decreases slowly until close to final failure. During the last 5% of fatigue life the stiffness drops suddenly. The results obtained for stress range control and strain range control are similar, despite the sudden loss of stiffness in the strain-controlled case. This behaviour is related to the rise of temperature in the specimen and the stress release. Tests by the authors show that this stress release at room temperature is very significant during the first 10 minutes after load application. Such behaviour can help explain the sudden drop of stiffness observed in the first stage of fatigue.

In Fig. 6 are plotted the variation of the stress range (in terms of $\Delta\sigma/\sigma_0$, where $\Delta\sigma$ is the stress range at any given moment of the test and σ_0 is the stress range at the start) against N/N_f for displacement control tests in the 0° laminate.

Similar plots were obtained for the other laminates. These curves are very close to E/E_0 versus N/N_f. This means that the variation of the stress range is mainly

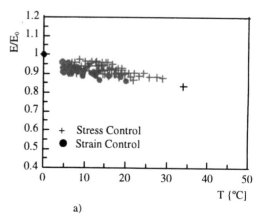

a)

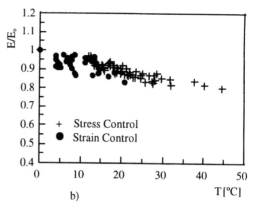

b)

Figure 7. E/E_0 against the rise of temperature ΔT. a) 0°; b) +45°/0°/-45°.

caused by stiffness decrease and the stress release is only a secondary effect.

Other important effects must also include in temperature and internal delaminations.

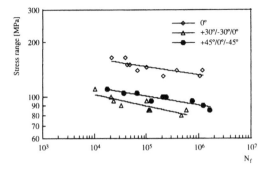

Figure 8. Influence of lay-up geometry on S-N curves. Stress controlled tests.

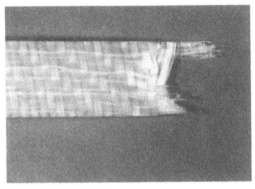

a)

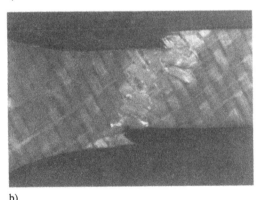

b)

Figure 9. Fatigue tests fracture specimens. a) 0°; b) +45°/0°/-45°.

Figs. 7a) and 7b) show plots of loss of stiffness E/E_0 versus the rise in temperature ΔT in the 0° and +45°/0°/-45° laminates, respectively.

For both laminates and both test conditions a nearly linear decrease of stiffness with the temperature was observed during the second stage. In the first fatigue stage the experimental points do not coincide with the

linear fitted curve. The most important cause of this difference must be the stress release observed in the first stage.

The results of the fatigue tests were plotted in terms of stress range versus the number of cycles to failure. Fig. 8 plots the fatigue results obtained under controlled stress mode.

The analysis of this figure shows that the fatigue strength of 0° laminates (where the fibre direction is always the same as the load) is much higher (1.5-1.8 times) than the other two laminates. This effect is related to the change of failure mechanism. In +45°/0°/45° and +30°/-30°/0° laminates the predominant fatigue mechanism is the debonding between the fibres and matrix caused by normal stresses. The fatigue strength obtained in +30°/ -30°/0° laminates is 10-15% lower than in +45°/0°/ -45 laminates which is a consequence of the higher normal stress component in the fibres with a 30° of inclination. The failure is observed in inclined planes by delamination along the fibre direction. In the 0° laminate longitudinal fibres predominantly absorb the normal stress and the failure is in transverse planes.

Fig. 9a) and b) shows the failure aspect observed in fatigue tests for the 0° and +30°/-30°/0° laminates, respectively.

The fracture aspect of fatigue failure is very similar to that observed in static failure (Fig. 2). The fatigue process in +45°/0°/-45° and +30°/-30°/0° laminates starts by debonding in transverse fibres. Long longitudinal delaminations can occur and finally broken longitudinal fibres were observed. The fatigue strength for +45°/0°/-45° and +30°/-30°/0° laminates is similar for low fatigue lives and only a little higher in +45°/0°/-45° for longer lives.

The fatigue strength obtained in displacement controlled tests is plotted in Fig. 10.

By comparison of the results plotted in Figures 8 and 10 only a small decrease in fatigue life is observed in all the three laminates in the case of displacement controlled tests. In spite of the stress release expected in the displacement-controlled mode, and the decrease in stress range caused by the loss of stiffness, the decrease

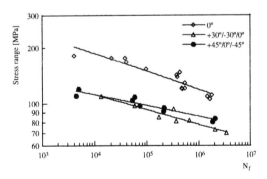

Figure 10. Influence of lay-up geometry on S-N curves. Displacement controlled tests.

in fatigue strength is very small (less than 5%). These results confirm that the effect of stress release was negligible.

4 CONCLUSIONS

1-Within the range of stain rate 0.333-0.0000333 s^{-1}, there was obtained only a small tendency to the increase of static properties with the strain rate. This tendency was more important for the failure stress in the 0° laminate. The ultimate strength for the +45°/0°/45° and +30°/-30°/0° laminates was only about 50% that obtained for the 0° laminate.

2-Similar damage mechanisms are observed in stress controlled tests and strain controlled tests both in 0° and in multi-angle +45°/0°/-45° and +30°/ -30°/0° laminates.

3-Only a small decrease in the fatigue strength (less than 5%) is observed in all three laminates in displacement controlled mode despite the decrease of stress range caused by the loss of stiffness during the test.

4-The fatigue strength is strongly influenced by the layer design. 0° laminate results have values 1.5-1.8 times higher than +45°/0°/-45° and +30°/-30°/0° laminates which exhibit similar fatigue strengths (although +30°/-30°/0° is 10-15% lower than the +45°/0°/-45 case).

5-The loss of stiffness (E/E_0) starts early in the fatigue process life. A sudden drop of E/E_0 (about 5%) is observed during the first 5% of the fatigue life and thereafter only a small stable decrease is observed until close to failure. During this period a linear relationship is observed between the loss of stiffness (E/E_0) and the temperature rise.

REFERENCES

Curtis P.T., "A Review of the Fatigue of Composite Materials", *Royal Aircraft Establishment, Technical Report 87031*, 1987.

Echtermeyer A.T., Engh B. and Buene L., "Lifetime and Young's Modules Changes of Glass/Phenolic and Glass/Polyester Composites under Fatigue", *Composites*, Vol. 26, N° 1, pp. 10-16, 1995.

Ferreira J.A.M., Costa J.D.M. and Richardson M.O.W., "Fatigue Behaviour of a Glass Fibre Reinforced Polypropylene Composite", Proc. 11th European Conference on Fracture, Poitiers, France, pp. 1653-1658, 1996.

Ferreira J.A.M., Costa J.D.M. and Richardson M.O.W., "Effect of Notch and Test Conditions on the Fatigue of a Glass-Fibre- Reinforced Polypropylene Composite", *Comp. Science and Techn.*, Vol. 57, pp. 1243-1248, 1997.

Hahn H.T. and Kim R.Y., "Fatigue Behaviour of Composite Laminate", *Journal of Comp. Materials*, Vol. 10, pp. 156-180, 1976.

Harper R.C. and Pugh J.H., "Thermoforming of Thermoplastic Matrix composites", *International Encyclopedia of Composites-Vol. 5*, pp. 496-530, S. M. Lee, ed., New York, VCH Publishers, Inc., 1991.

Harris B., Reiter H., Adam T., Dickson R.F. and Fernando G., "Fatigue Behaviour of Carbon Fibre Reinforced Plastics", Composites, Vol. 21, N° 3, pp. 232-242, 1990.

Hou M., Lin Y. and Mai Y., "Advances in Processing of Continuous Fibre Reinforced Composites with Thermoplastic matrix", *Plastics, Rubber and Composite Processing and Applications*, Vol. 23, N° 5, pp. 279-293, 1995.

Joseph E. and Perreux D., "Fatigue Behaviour of Glass-Fibre/Epoxy-Matrix Filament-Wound Pipes: Tension Loading Tests and Results", *Comp. Sci. Tech.*, Vol. 52, pp. 469-480, 1994.

Pink E. and Campbell J.D., "Deformation Characteristics of Reinforced Epoxy Resins, Part 1, The Mechanical Properties", *J. Mat. Sci.*, Vol. 9, pp. 658-664, 1974.

Sims G.D. and Gladman D.G., "Effect of Test Conditions on the Fatigue Strength of a Glass - Fabric Laminate: Part A - Frequency", *Plast. and Rubber: Mat and Appl.*, pp. 41-48, 1978.

Sims G.D. and Gladman D.G., "Effect of Test Conditions on the Fatigue Strength of a Glass - Fabric Laminate: Part B - Specimen Condition", *Plast. and Rubber: Mat and Appl*, pp. 122-128, 1980.

Experimental Techniques and Design In Composite Materials 4, Found (Ed.)
© *2002 Swets & Zeitlinger, Lisse, ISBN 90 5809 370 0*

Fatigue damage of carbon fibre reinforced laminates under two-stage loading

M.S. Found
Department of Mechanical Engineering, University of Sheffield, UK

M. Quaresimin
Department of Management and Engineering, University of Padova, Vicenza, Italy

ABSTRACT: The cumulative fatigue damage of a [0/90,±452,0/90]s quasi-isotropic carbon fibre-epoxy laminate was analysed by testing the material under two-stage loading conditions. Several tension fatigue tests were carried out at stress ratio R=0.05 and the failure mechanisms were investigated by comparing x-rays taken after each loading stage. The results show that a low-to-high loading sequence is more damaging than a high-to-low one and also confirm that the Palmgren-Miner linear damage rule may no longer be valid for this kind of material, as reported in the literature. A brief review of the models used to predict the cumulative fatigue damage is presented and the new results, in terms of fatigue strength and damage growth, are compared with those in the literature for carbon-epoxy laminates.

1 INTRODUCTION

A great deal of work has been carried out in recent years to improve the understanding on the fatigue behaviour of composite materials at constant amplitude. A large amount of experimental data are available which may be useful in the fatigue design of structural members. Several models have been developed for the prediction of the fatigue life and strength of these materials, even if the problem is still far from complete due to the lack of valid design rules for the wide range of fibres, matrices and lay-up available.

The situation is very different for the more realistic case of variable amplitude loading, since a structural member is usually subjected to a load spectrum far from the constant amplitude case. For this situation there are very limited experimental data (1-18), some of which (13-18) are on glass reinforced plastics.

A common theme in the available data, both for carbon and glass reinforced materials, is that damage accumulation follows a non-linear law. The assessment of the fatigue life with the Palmgren-Miner linear damage rule (19, 20) can in fact give very dangerous results either for simple (2, 4, 7, 8, 13, 16, 18) or complex block and real spectrum loading conditions (1, 3, 6, 9, 12). It is worth noting, however, that in few cases the linear damage rule turned out effective (9, 10, 12) (in Refs. 9 and 12 only for all-tension or all-compression block loading). Other evidence is that, for two-stage loading, a low-to-high loading sequence is generally more damaging than a high-to-low sequence (2, 4, 5, 7, 8, 13, 14, 18).

Nevertheless, there is an urgent need of experimental data and design rules enabling the designer to predict the fatigue life of composite structures under multi-stress levels or random loading conditions. With a safe-life design approach, the lack of sound life prediction methodologies implies the adoption of large safety factors, resulting in an increase of weight, and cost, of the composite structure. This is unacceptable if we consider that the high specific strength and stiffness values are probably the most important reason for using advanced composite laminates.

This paper reports the results of an experimental programme carried out on a plain quasi-isotropic carbon fibre-epoxy laminate with the aim of obtaining preliminary fatigue data useful for the assessment of a model for the cumulative fatigue damage of the material under variable amplitude loading conditions.

2 CUMULATIVE DAMAGE MODELS

The accumulation of damage for variable amplitude fatigue have been analysed by many researchers and quite a large number of fatigue damage models have been developed either from experimental evidence or from theoretical analysis. Extensive reviews of the cumulative damage models have already been presented (Hwang and Han 1986, 1989) and the aim of this brief discussion is to review some of the most used models and to describe those that will be used in the following analyses.

The two principal approaches to the life prediction are the residual life method and the residual strength (or stiffness) method. Both these methods relate failure to a specific value of a physical quantity: in fact, the former predicts failure when the residual life equates to zero while in the latter failure occurs when the residual strength of the material decreases until the maximum applied fatigue stress is reached. This intrinsic similarity has been considered by Hashin (1985) who demonstrated that the two methods lead to substantially equivalent results.

These methods and the relevant models enable life prediction on the basis of previous fatigue test programmes at constant or, less frequently, variable amplitude. They therefore require a large amount of experimental data for each material, lay-up and loading condition of interest. Although the need for an experimental fitting can be seen as a disadvantage, the above methods seem to be, at present, the more promising for formulating sound rules for design. In fact from a research point of view, it is difficult at present to imagine the development of models of general validity.

Let us consider some of the models for both the residual strength and the residual life method. The first model developed for evaluating the cumulative fatigue damage in metallic materials is the well known Palmgren-Mincr (1924, 1945) linear rule which states that failure occurs when:

$$D = \sum_1^m \frac{n_i}{N_i} = 1 \qquad (1)$$

where D is a damage parameter

(n/N) is the fractional life.

This relationship is often used as a means of comparison also for composite materials. The first residual strength model for composite materials seems to be due to Broutman and Sahu (1972) who found that the residual strength is related to the cumulative fatigue damage and is also a linear function of the fractional life spent at a given stress level. They proposed the following model for the cumulative fatigue life under two-stage loading:

$$\frac{(\sigma_{uts} - \sigma_1)}{(\sigma_{uts} - \sigma_2)} \cdot \frac{n_1}{N_1} + \frac{n_2}{N_2} = 1 \qquad (2)$$

where σ_{uts} is the tensile static strength

σ_1, σ_2	are the maximum stress levels in the first and second stages
n_1, n_n	are the number of cycles for each stage
N_1, N_2	are the mean lives for each stage

Similar models have been proposed by Yang and Jones (1980, 1981, 1983), with the major and improvement being to incorporate statistical variations within the model. Whitworth (1990) presented a stress dependent residual stiffness model which modified the Miner summation accounting for the stiffness degradation.

Considering a residual life model, that presented by Marco and Starkey (1954) suggests that the damage may be represented by the simple non-linear equation:

$$d = \left(\frac{n}{N}\right)^\alpha \qquad (3)$$

where the coefficient α is a function of the applied stress, to be determined on the basis of experimental results.

The failure condition for two-stage loading can be written in the form of an exponential damage rule as:

$$D = \sum_1^m d_i = \left(\frac{n_1}{N_1}\right)^{\alpha_1} + \left(\frac{n_2}{N_2}\right)^{\alpha_2} = 1 \qquad (4)$$

On the assumption that fatigue damage represented by the S-N curve added logarithmically, Hashin and Rotem (1978) modified the previous exponential damage law and proposed the equation:

$$\left(\frac{n_1}{N_1}\right)^{(1-s_2)/(1-s_1)} + \left(\frac{n_2}{N_2}\right) = 1 \qquad (5)$$

where the exponent $(1-s_2)/(1-s_1)$ is no longer evaluated experimentally, but depends on both the applied stress levels and the static strength of the material ($s_1 = \sigma_1/\sigma_{uts}$, $s_2 = \sigma_2/\sigma_{uts}$).

The life predictions of the models described above will be compared with our experimental results presented for two-stage loading. However, for a more exhaustive analysis of the available cumulative damage models we refer the reader to the reviews by Hwang and Han (1986, 1989).

3   EXPERIMENTAL DETAILS

This study was undertaken on a 8 ply, five-harness satin weave fabric pre-impregnated with epoxy resin Fibredux 914C-833-40% (Hexcel Composites 914 epoxy resin matrix with 40% nominal resin content by weight, Toray 300 carbon fibre with 3000 filament per tow). The laminates were laid up as $[0/90, \pm 45_2, 0/90]_s$ and autoclave moulded by Hurel-Dubois UK. The specimens were cut into blanks nominally of 250x20x2.25 mm and the ends reinforced with 50 mm glass fibre laminate tabs bonded with Redux 403 Ciba-Geigy epoxy structural adhesive. The static strength of the material was determined by Oxley (1991) giving $\sigma_{uts} = 422$ Mpa C. of V.=0.045).

Tension fatigue tests were carried on a 30 kN loading frame driven by a hydraulic pump, under load control with a sinusoidal wave form and a stress ratio R=0.05. The frequency was about 3 Hz, which give a negligible temperature rise during the tests.

Some preliminary constant amplitude fatigue tests were carried out to complete the data already reported (Oxley 1991) for the same material. All the data were analysed together to evaluate the reference fatigue curve at constant amplitude.

For the analysis of the cumulative fatigue damage several specimens were tested under two-stage loading conditions, namely low-to-high loading and high-to-low loading, always at R=0.05. The testing procedure consisted of two blocks of fatigue cycles, the first one at a specified fatigue life ratio and the second one, at a different stress level, until failure of the specimen occurred. The stress levels were chosen as follow: maximum stress in the low stress stage σ_L=315 MPa and maximum stress in the high stress stage σ_H=340 MPa, which gave a mean fatigue life at constant amplitude of N_L=115150 and N_H=8800 cycles, respectively.

After the first stage of loading and after failure a damage assessment of the specimens was carried out by means of enhanced x-radiography. For the enhancement, a solution of zinc iodine was used and the radiographs were taken by using Structurix D4 AGFA x-ray film in a HP Faxitron x-ray Cabinet, with the following set-up: 25 kV, 3 mA and 120 sec.

4 FATIGUE RESULTS

The experimental results of the two-stage loading tests are shown in Figure 1, where the remaining fractional life of the second stage is plotted as a function of initial fractional life spent in the first stage. Some data were not considered in the analysis due to their anomalous behaviour with respect to the others of the same group. In particular, for the high-to-low datum the quick failure after the first stage was due to the almost complete delamination of one of the external layers and the subsequent presence of out-of-plane stress. For the low-to-high data not accounted for, the reason of the longer lives may be due to the large scatter at high stress level, also recorded in the constant amplitude fatigue tests.

In Figure 1 also plotted is the life prediction given by the linear Palmgren-Miner rule (1924, 1945) and it can be seen that the damage summation for the specimens does not follow the linear rule but instead seems to follow a highly non-linear damage rule, as can be expected from the literature review previously reported.

Hence, the data for both low-to-high and high-to-low tests are fitted with the damage curves provided by equation (4), which are also plotted in Figure 1. The numerical best fit of the experimental data carried out for evaluating the exponents gave the following results:

Low-to-high loading curve: $\alpha_{1,LH} = 0.815$
$\alpha_{2,LH} = 0.525$
High-to-how loading curve: $\alpha_{1,HL} = 1.087$
$\alpha_{2,HL} = 1.641$

In Figure 2 the life prediction of the exponential law model fitted on the experimental data is compared with those given by the Broutman-Sahu (1972) and the Hashin-Rotem (1978) models previously presented. For the evaluation of the coefficients for both models the ratios between the applied stresses and the static strength of the material were used. The best agreement between the models and the experimental data have been obtained for the exponential law model, and this could seem obvious since it is the results of a numerical best fit. However it is interesting to note how both the Broutman-Sahu and Hashin-Rotem models strongly overestimate the fatigue life for the low-to-high condition, even if their coefficients were calibrated for the material and the test condition under examination. Further remarks can be made about the Broutman-Sahu

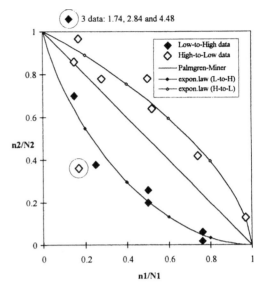

Figure 1. Fatigue data compared with linear and exponential models.

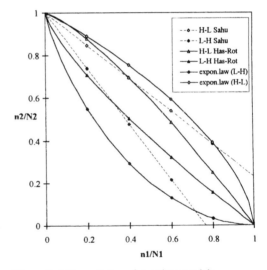

Figure 2. Life predictions for various models.

model in that the life prediction is always made on the basis of a linear, although modified, rule and therefore the model will not give accurate prediction in all the cases in which the trend of the experimental data is highly non-linear. Moreover it does not agree with the condition to give residual life equal to zero in the second stage when all the life is spent in the first stage (n_1N_1 =1).

It is also worthwhile to note the strong deviation from the linearity of the experimental data even though the difference in the low and high stress levels applied during the fatigue tests was not high. In fact it has been found by Broutman and Sahu (1972), for glass reinforced laminates, that the larger is the difference between the stress levels in the first and second stage, the greater is the decrease for the low-to-high stage and the increase for the high-to-low stage of the linear damage sum with respect to unity. This fact is well reproduced by their model and by that of Hashin and Rotem (1978).

The influence of the loading sequence in the fatigue life of the material is quite evident from the analysis of Figures 1 and 2: the low-to-high loading condition turned out to be strongly more damaging than the high-to-low, as we expected. This behaviour has already been recognised in many works (2, 4, 5, 7, 8, 13, 14, 18), as reported in the introduction, and its explanation could be related to the peculiarity of the fatigue damage mechanisms of composite materials. It is well known that under constant amplitude loading fatigue failure in composites occurs as a result of accumulation of many micro-cracks rather than by the propagation of one dominant crack. At low stress levels, because of the reduced fraction of the applied load carried by the fibres, the damage mainly involves the matrix with growth of microcracks throughout the specimen, which can induce a rapid failure in the following high stress level stage. At high stress levels the load is carried mostly by the fibres so, although the damage may be more severe, it is more concentrated at the fibre-matrix interface and the development and the growth of microcracks through the matrix is more difficult, resulting in damage of the specimen which is lower with respect to that at the low stress level for the same fraction of life. The proposed explanation can find some experimental validation in the analysis of the damage growth for the material presented in the following section and in some works in the literature which will be analysed later.

Further investigation is however needed in the zone of low fractional life for the low-to-high loading sequence, since the linear damage sum for some specimens gave values much higher than unity (see Figure 1). This may be due to the intrinsic scatter of the fatigue data or to a favourable relaxation of the manufacturing induced cure stresses by the low stress applied for a few cycles but also to a change of the failure mechanism.

For low fractional life spent at the low stress level it appears that the behaviour is similar to that of metallic materials for which a previous fatigue cycling at a low stress level can lead to improve the fatigue performance (Cazuad et al. 1969, Frost et al. 1974) A damage summation greater than unity for a part of the fractional life and lower for the remaining has been observed also for low-cycle fatigue at high temperature of steel for both low-to-high and high-to-low stage loading (Miller and Gardiner 1977).

However, the assumption made to not consider in the definition of the damage curve the data with damage summation higher than unity is conservative from the design point of view and therefore it can considered acceptable.

5 ANALYSIS OF DAMAGE

The growth of the damage during the fatigue life of the material was investigated by means of x-radiography and also through microscopic observation, after the first stage of loading and after failure. At the end of the first stage there were no visible signs of any damage on the surface of the specimens for both loading conditions. A microscopic analysis of the edges of the specimens showed evidence of the presence of delamination and transverse matrix cracking, mostly for the specimens with fractional life higher than 0.5 and loaded at the low stress level. On the contrary, after failure the high-to-low loaded specimens presented, generally, a more damaged surface with extensive delaminations and surface cracks.

The x-radiography allow, however, a more accurate analysis of the growth of the damage and the results are presented in Figure 3 where the delaminated areas, normalised with respect to the total area of the specimen, are plotted as a function of the fractional life.

After the first stage, only the low-to-high loaded specimens were found to show evidence of delamination, as also shown in Figure 4 where are compared the x-radiographs, after the first stage, for two specimens loaded for the same fractional life (about 0.75) respectively at low and high stress level.

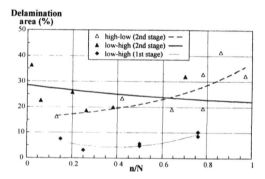

Figure 3. Normalised delamination areas for fatigue tests.

Figure 4. X-radiographs at $n_1/N_1 \sim 0.75$ for specimens at a) low stress and b) high stress.

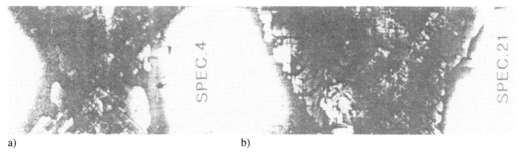

a)   b)

Figure 5. X-radiographs of failure zone for a) low-high and b) high-low loading.

After failure, in spite of the large scatter of the results, the trend lines of the delamination area seem to be independent of the sequence of loading being rather than being related to the fraction of life spent at low stress. However, it has to be considered that for the evaluation of the trend line for the low-to-high sequence the data with fractional life greater than unity were not considered and also that by using the x-radiography it is only possible to obtain qualitative information because this technique gives results averaged through the thickness of the sample.

After failure the damage due to the fatigue loading appeared to be similar: both the low-to-high and high-to-low loaded specimens presented wide delamination areas involving in most cases the 45° plies, as can be seen in Figure 5. The onset of almost all the failures was due to a 45° split emanating from an area of delamination and growing completely across the specimen width until failure occurred, with a characteristic x- shaped delamination area and a strong delamination of the 45° plies. Sometimes the delamination still involved the 45° plies but the specimen failed due to a 90° split (Y-shaped delamination area).

An overall look at the results, obtained in terms of both fatigue life and fatigue damage, highlights the influence of the loading sequence of the fatigue life of this composite material but also seem to indicate that the higher the fraction of life spent at low stress level, the higher is the damage in terms of fatigue life, delamination and associated properties like residual stiffness. As a confirmation of this, Figure 6 compares the stiffness trend, normalised with respect to initial values, for low-to-high and high-to-low tests. In the low-to-high sequence, a greater stiffness decrease has been noticed in the first stage, while in the high-to-low sequence the second stage is the more damaging in terms of stiffness loss. In both cases, at low stress level, the stiffiless loss is about three times that at high stress level and also the stiffness loss gradient is greater. Further evidence of the detrimental influence of the fraction of life spent at low stress level can be found in some works in the literature.

Investigating the behaviour of a notched quasi-isotropic $[0/90/\pm 45]_2 s$ T800/5208 graphite-epoxy (Jen et al. 1994) found, for constant amplitude tests, a greater

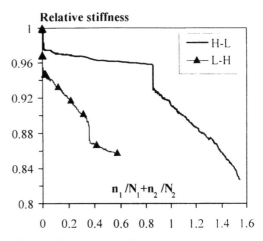

Figure 6. Reduction of stiffness with damage.

delamination area for low stress level than for high stress while in the latter case the damage was more concentrated through the thickness. During two-stage fatigue loading on the same material, they also found for the low-to high loading sequence a greater extent of the damage associated with a higher decrease of the normalised stiffness, at the same fractional life.

The influence of the stress level on the damage accumulation has been investigate for plain and notched carbon-epoxy laminate under different test condition at constant amplitude (Razvan et al. 1988). For AS4/1808 $[0/45/0/-45]s_4$ notched specimens tested under tension-compression fatigue load it was found that there was a higher damage level under low stress than under high stress at the same fraction of life in terms of both delamination area and normalised stiffiless reduction. Even if it was suggested that the damage modes differ, primarily in extent and not in their fundamental nature under the different load levels, more matrix cracking and delamination was found in the low level case, at the same fraction of life and the fatigue fracture appeared more localised for specimens under a high load condition than under a low condition, resulting in a more concentrated damage.

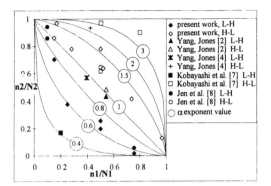

Figure 7. Influence of exponent of exponential model.

Poursartip and Beaumont (1986) analysed the extent of the normalised delamination area considered as a damage parameter during fatigue tests at constant and variable amplitude on a (45/90/-45/0)s XAS/914 carbon-epoxy laminate. For a given loading history they found that the larger the number of cycles at low stress level the greater is the extent of the damage rate, particularly, if less than 10% of cycles at high stress level are present in the loading history giving an increase of about three times of the damage rate.

6 COMPARISON WITH THE LITERATURE

The problem of the two-stage fatigue loading of composite materials has been considered for over twenty years ago. In spite of this, at present, there are very few works with experimental data on two-stage fatigue loading. The data of the different works are compared with ours in Figure 7, where are also drawn the life predictions given by equation (4) for different values of the exponent (for simplicity we equated the two exponents, $\alpha_1=\alpha_2=\alpha$). Yang and Jones (1980, 1982) carried out the two-stage fatigue tests with the aim of investigating the statistical scatter of the results at a fixed fraction of life, therefore in Figure 7 are plotted only the mean values of those results.

The aim of the present comparison is to analyse the influence of material system, lay-up and difference in stress level on the two-stage fatigue behaviour of carbon-epoxy laminates. Although almost all the works refer to quasi-isotropic laminates with about the same fibre volume fraction, it is quite evident in Figure 7 the scatter of the results. Only in one of the Yang and Jones works (1980) the lay-up is such to induce a matrix dominated behaviour of the material and this could probably justify the damage summation close to unity, as occurs also for the data of Jen et al. (1994). In the latter case, however, the material is quasi-isotropic and the presence of the notch could have reduced the sensitivity of the material to the higher damage due to the low level stress.

Eventually, even the difference in the stress level between first and second stage seems to have no clear influence on the deviation of the damage sum from the linearity, as can be recognised from the comparison among our data and those reported by Kobayashi et al. (1987) and Jen (1994). This is quite in contrast with that reported by Broutman and Sahu (1972) for glass-reinforced laminates; in which they found that the deviation from the linearity of the damage summation in much higher as the difference in stress level between the first and second stage increases.

An overall look to all the results presented in Figure 7 can however confirm the previously mentioned influence of the loading sequence: that is the low-to-high data have normally a damage summation lower than unity, while those for high-to-low loading sequence have a damage summation greater than unity. Unfortunately the high variation of the data does not allow us to identify a damage curve which could, as a first approximation, be used to generally represent the behaviour described.

7 CONCLUSIONS

The fatigue behaviour of a plain $[0/90,\pm45_2,0/90]_s$ quasi-isotropic carbon fibre-epoxy laminate under two-stage loading was investigated. The main conclusions can be summarised as follows:

The loading sequence can strongly influence the fatigue life of this material: a low-to-high sequence turned out to be much more damage than a high-to-low sequence, in terms of both fatigue life and delamination growth.

The accumulation of damage follows a non-linear rule; the prediction of fatigue life by means of a linear damage rule can dangerously overestimate the fatigue life in the low-to-high sequence while it can produce an underestimation of the life for the high-to-low loading condition. Different damage models have been used to fit the experimental results and the best agreement was found with an exponential damage law.

Further experimental investigations are required for a better understanding of the behaviour of this material for low fractional life in the first loading stage and for investigating the effects of the two-stage loading under different stress ratios.

A comparison among the results obtained in the present work and those available in literature put in evidence the impossibility, at least at present, to relate the two-stage fatigue behaviour with material properties and level of the applied stress due to the lack of a statistically meaningful number of data. Moreover the large variation of the experimental data require further and deeper investigations in an attempt to find a general relationship to predict the fatigue life first under two-stage loading condition and later under real spectrum loading.

8 ACKNOWLEDGEMENTS

One of the authors (M.Q.) wish to thank the Italian National Research Council (CNR) for the financial support obtained in the frame of the Programme for International Exchange and Short Mobility 1996-1997

REFERENCES

Schutz D. and Gerharz J.J. 1977. Fatigue strength of a fibre-reinforced material. *Composites*, **8**, no.4, 245-50.

Yang J.M. and Jones D.L. 1980. Effect of Load Sequence on the Statistical Fatigue of Composites. *AIAA Journal*, **18**, no.12, (Article no.79-0760R), 1525-1531.

Philipps E.P. 1981. Effect of truncation of a predominantly compression load spectrum on the life of a notched graphite/epoxy laminate. In: Fatigue of fibrous *composite materials*, ASTM STP 723 American. Society for Testing and Materials, Philadelphia, 197-212.

Yang J.N. and Jones D.L. 1982. Fatigue of graphite/epoxy $(0,90,45,-45)_s$ laminates under dual stress levels. *Comp. Tech. Rev.*, **4**, no.3, 63-70.

Yang J.N. and Jones D.L. 1983. Load sequence effects on graphite/epoxy $(\pm 35)_{2s}$ laminates. In: *Long-term Behavior of Composite*, ASTM STP 813, American Society for Testing and Materials, Philadelphia, 170-188

Poursartip A and Beaumont P.W.R. 1986. The fatigue damage mechanics of a carbon fibre composite laminate: II - Life predictions. *Comp. Sci. Tech*, 25, no.4, 283-299.

Kobayashi A., Othani N. and Choi K.B. 1987. Microscopic inner materials damage due to two level fatigue in composites. *Proceedings of 6th International Conference on Composite Materials* (Edited by F. L. Matthews, N. C. K. Buskell, J. M. Hodgkinson and J. Morton), Elsevier Applied Science, 3, pp.3.176-3.184.

Jen M.H.R., Kau Y.S. and Wu I.C. 1994. Fatigue damage in a centrally notched composite laminate due to two-step spectrum loading. *Int. J Fatigue*, 16, no.3, 193-201.

Adam T., Gathercole N., Reiter H. and Harris B. 1994. Life prediction for fatigue of T800/5425 carbon-fibre composites: II. Variable amplitude loading. *Int. J Fatigue*, **16**, no.3, 533-547.

Noguchi H., Kim Y.H. and Nisitani H. 1995. On the cumulative fatigue damage in short carbon-fiber reinforced poly-ether-ether-ketone. *Engng. Fract. Mech.*, **51**, no.5, 457-468.

Found M.S. and Kanyanga S.B. 1996. The influence of two-stage loading on the longitudinal splitting of unidirectional carbon-epoxy laminates. *Fatigue Fract. Engng. Mater. Struct.*, **19**, no.1, 65-74.

Harris B., Gathercole N., Reiter H. and Adam T. 1997. Fatigue of carbon-fibre reinforced plastics under block-loading conditions. *Composites A*, **28A**, 327-337.

Broutman L.J. and Sahu S. 1972. A new theory to predict cumulative fatigue damage in fibreglass reinforced plastics. In: *Composite Materials: Testing and Design*, ASTM STP 497, American Society for Testing and Materials, Philadelphia, 170-188

Tanimoto T. and Amijima S. 1981. Fatigue life and its reliability of FRP under multi-step loading, *Composite Materials: Mechanics, mechanical properties and fabrication.* Proceedings of the Japan-U.S. Conference, (Edited by K. Kawata and T. Akasaka, Tokio)

Hwang W. and Han K.S. 1989. Fatigue of composite materials - Damage model and life prediction, In: *Composite Materials: Fatigue and Fracture, 2nd Volume*, ASTM STP 1012, American Society for Testing and Materials, Philadelphia, 87-102

Jessen S.M. and Plumtree A. 1991. Continuum damage mechanics applied to cyclic behaviour of a glass fibre composite pultrusion, *Composites*, **22**, no.3, 181-190.

Kallmayer A.R. and Stephens R.I. 1995. Constant and variable amplitude fatigue behaviour and modelling of an SRIM polymer matrix composite. *J Comp. Mat.*, **29**, no.12, 1621-1648.

Otani N. and Song D.Y. 1997 Fatigue life prediction of composite under two-step loading, *J. Mater. Sci.*, **32**, no.32, 755-760.

Palmgren A. 1924. Die Lebensdauer von Kugellagern. *Zeitschrift des Vereins Deutscher Ingenieure*, **68**, 339-341.

Miner M.A. 1945. Cumulative damage in fatigue. *J. App. Mech.*, **12**, A159-A164.

Hwang W. and Han K.S. 1986. Cumulative damage models and multi-stress fatigue life prediction, *J Comp. Mat.*, **20**, no.20, 25-153.

Hashin Z. 1985. Cumulative damage theory for composite materials: residual life and residual strength methods. *Comp. Sci. Tech.*, **23**,1-19.

Yang J.N. and Jones D.L. 1981. Load sequence effects on the fatigue of unnotched composite materials. In: *Fatigue of Fibrous Composite Materials*, ASTM STP 723, American Society for Testing and Materials, Philadelphia, 213-232

Whitworth H.A. 1990. Cumulative damage in composites. *Trans. ASME, J Engng. Mat. Tech.*, **112**, 358-361.

Marco S.M. and Starkey W.L. 1954. A concept of fatigue damage. *Trans. ASME*, **76**, 627.

Hashin Z. and Rotem A. 1978. A cumulative damage theory for fatigue life prediction. *Mat. Sci. Engng.*, **34**, 147-160

Oxley M. 1991. The effect of low velocity impact damage on the performance of a woven CFRP. *Ph.D. Thesis*, University of Sheffield.

Cazuad R., Pomey G., Wabbe P. and Janssen Ch. 1969. La fatigue des mètaux, Dunod, Paris (in French).

Frost N.E., Marsh K.J. and Pook L.P. 1974. Metal fatigue, Clarendon Press, Oxford.

Miller K.J. and Gardiner T. 1977. High-temperature cumulative damage for stage I crack growth. *J. Strain Anal.*, **12**, no.4, 253-261.

Razvan A., Bakis C.E., Wagnecz L. and Reifsnider K.L. 1988. Influence of cyclic load amplitude on damage accumulation and fracture of composite laminates. *J. Comp. Tech. Res.*, **10**, no.1, 3-10.

Section 2: *Test Methods*

Experimental Techniques and Design In Composite Materials 4, Found (Ed.)
© 2002 Swets & Zeitlinger, Lisse, ISBN 90 5809 370 0

Preparation procedure for MMC$_p$ fatigue test specimens

A. De Iorio, D. Ianniello, R. Iannuzzi & F. Penta
Dipartimento di Progettazione e Gestione Industriale, University Federico II, Naples, Italy

G. Florio
Istituto Sperimentale Ferrovie dello Stato, Rome, Italy

G. Scalabrino
Dipartimento di Energetica e Termofluidodinamica Applicata, University Federico II, Naples, Italy

ABSTRACT: In this paper the authors investigate a suitable procedure for the traditional machining of particulate MMC, in order to make round specimens for low cycle fatigue tests. Standards actually in use for unreinforced metals (ASTM, BSI, etc.) do not provide information about machining round specimens for composite materials. An experimental investigation was conducted for a Ti6AI4V/TiCp in order to estimate the cutting parameters for turning (depth of cut, cutting speed and feed) that minimise material damage and residual strain due to machining, that could compromise the fatigue tests data.

1 INTRODUCTION

Powder Metallurgy (PM) has gained a very important role in industrial applications. It is a suitable way to produce mechanical components with very complex shape that would be hard to realise by traditional machining. Although it is a near-net shape process, the inability to produce certain geometrical figures (such as holes, threads, etc.) and to obtain surfaces with a high finished grade, frequently necessitates some machining (Chandler 1989).

The target of the present work is to evaluate the influence of cutting parameters on the fatigue behaviour of MMC components made by PM.

2 MATERIAL

The investigated material is a metal matrix composite consisting of a titanium alloy matrix (Ti6AI4V) reinforced by TiC particles (10% wt). It was produced by Dynamet Technology for aero-engine applications and its commercial name is CermeTi-C-10. This material was produced in the form of cylindrical bars obtained by CHIP (cold isostatic pressuring, vacuum sintering and hot isostatic pressing), that was intended to make specimens for low cycle fatigue tests.

Before machining, metallographic examinations were made about the microstructure of the material; the same examinations were made after machining on the cross sections and on the lateral surfaces of the workpieces in order to evaluate the influence of cutting parameters on the microstructure and the surface integrity of CermeTi-C-10.

The structure of the CermeTi-C-10 "as CHIPed" is shown in Figures 1, 2. The matrix has a typical microstructure of α + β-Ti alloys; the α-phase is characterised by equiaxed grains and the β-phase has an intergranular acicular morphology. The carbides are non-uniformly distributed in the matrix; there are some areas where they are clustered and form agglomerates whose size can reach about 10 microns.

Voids (or porosity) are clearly shown in Figure 2; they are situated near the matrix-carbides interfaces but sometimes they can be found inside the TiC particles too (see Fig. 9).

3 CUTTING TESTS

3.1 Cutting parameters

It is well known that P M materials have poor machinability in comparison with wrought or cast metals. This

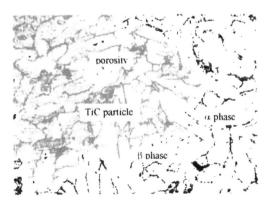

Figure 1. CermeTi-C-10 microstructure (500X).

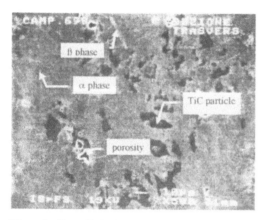

Figure 2. CermeTi-C-10 microstructure (SEM).

is due to the presence of porosity which affects the strength of materials, its thermal conductivity and cutting conditions.

Porosity reduces mechanical strength and this often results in better machinability; but its presence causes a discontinuous contact between the workpiece and the cutting edge of the tool, leading to both a reduction of tool life and a worse surface finish compared to cast metals machined in the same conditions (Chandler 1989). CermeTi-C-10 cutting conditions are further penalized by the presence of hard particles in the matrix such as titanium carbides.

Moreover, titanium is characterized by a value of thermal conductivity that is lower than the value for iron alloys (about one tenth) and the porosity causes a further decrease in this physical property referred to the composite. This is disadvantageous for the machinability of CermeTi-C-10 because the heat produced during cutting (due to plastic strain and the friction between tool edge and workpiece) diffuses slowly in the bulk material and its mechanical properties do not decrease enough. This results in a higher temperature in the cutting area that should cause oxidation and/or chemical reaction between the generated surface and the environment (i.e. cutting fluid, tool material, etc.). A direct consequence of the aforementioned problems should be a chemical-physical alteration of the material close to the surface.

All the above mentioned problems suggest how the cutting parameters could influence the surface finish and the microstructure of the material close to the surface. In order to evaluate these effects, turning tests were made using a traditional lathe and coated (TiN) carbide tools having a suitable geometry for titanium alloys (0° back rake angle, 5° side rake angle). Cutting fluid was used during the turning in order to avoid local over-heating in the cutting area.

The cutting parameters range were established on the basis of recommendations, found in literature, referring to the machinability of titanium (Donachie, Jr 1989)

because better data were not available for Cerme-Ti-C-10. After some preliminary tests, the following values were established:
1. three values for the depth of cut, p: 1.5; 0.5; 0.075mm;
2. two values for the feed, f: 0.13; 0.065 mm/rev;
3. for cutting speeds, Vt, a variable range between 7.2 and 19.25 ml/min resulted from the combination of mandrel angular velocity and the diametral size of the workpiece. The workplan is illustrated in Table 1.

Table 1. Cutting parameter values

Test n°	f (mm/rev)	p (mm)	V_t (m/min)
58 A	0.065	1.5	19.250
61 B	0.065	0.075	17.849
27 A	0.067	0.075	17.797
56 B	0.13	0.5	17.449
62 B	0.13	0.075	17.376
57 A	0.065	0.5	16.976
66 B	0.13	0.75	16.944
62 A	0.065	0.5	15.965
66 A	0.13	1.5	15.681
61 A	0.065	1.5	13.997
56 A	0.13	1.5	12.808
70 B	0.065	0.075	11.714
57 B	0.065	0.5	10.491
30 A	0.065	0.5	10.143
30 B	0.13	0.5	10.120
14 B	0.13	0.075	10.120
14 A	0.065	0.075	10.098
18 B	0.13	1.5	9.935
18 A	0.065	1.5	9.920
70 A	0.065	0.075	8.572
69 B	0.065	0.075	8.545
69 A	0.13	0.075	8.508
59 B	0.065	0.5	8.022
67 B	0.13	0.5	8.011
27 B	0.065	0.5	7.445
67 A	0.13	1.5	7.327
59 A	0.067	1.5	7.194

During tests, temperature measurements were made in the cutting zone in order to evaluate if overheating should be overcome. The method and the instrumentation chosen for this purpose will be illustrated in the following section.

3.2 *Temperature measurement*

The choice of a suitable method for temperature measurement in this particular application is influenced by a great number of factors.

First of all, we must measure temperature on a moving surface of the workpiece (cutting area); moreover the presence of a cutting fluid introduces some other difficulties because the refrigerant acts as a thermal insulator and one can measure the fluid temperature believing it to be that of the workpiece.

Furthermore, it is desirable to have instrumentation with a very low time constant since we are going to measure continuously the temperature at the cut point, so the thermometer must reach the thermal equilibrium immediately. This is very important considering that the material of the workpiece is characterised by a very low thermal conductivity.

In order to minimise the foregoing problems, the most suitable method for temperature measurement is a non-contact technique (Cascetta & Vigo, 1988). These techniques introduce other metrological problems.

The first of these is the determination of the material emissivity. In order to overcome this issue, a thermostatic bath (realised with the same cutting fluid) is used. In this way the emissivity is determined under thermal equilibrium conditions. Nevertheless, during cutting operations the quantity of the fluid on the workpiece is continuously variable and as a result the emissivity is not a constant.

A second important problem is the meter target. While in theory the temperature would be taken only at the cut point (target), in practice the meters give an average temperature in a small area around it including some parts of the workpiece surface not under consideration.

All the above mentioned problems cause sources of uncertainties that must be assessed in order to evaluate if the temperature rise can affect the material microstructure.

In order to estimate the temperature in the cutting area an IR thermopile radiation thermometer was used. This is a non-contact meter that measures the temperature of a body by detection of radiant thermal energy emitted by the body in a suitable wave range (8 ÷ 14 μm). Finally, in order to minimise other sources of uncertainty (such as the relative motion of the workpiece, the influence of the surrounding environment, etc.) a purpose built structure was designed and realised to support the thermometer and to point the measurement target.

Cutting conditions during tests were not able to produce over-heating in the cutting zone, so they did not affect the microstructure. Table 2 indicates the temperatures corresponding to extreme values of cutting parameters; the highest value was lower than 60°C.

Table 2. Temperature Measurements

Test n°	f (mm/rev)	p (mm)	V_t (m/min)	T (°C)
58 A	0.065	1.5	19.250	57
66 A	0.13	1.5	15.681	40
56 B	0.13	0.5	17.449	42
59 A	0.065	1.5	7.194	32
27 B	0.065	0.5	7.445	30
69 B	0.065	0.075	8.545	28

All the temperatures were measured at a distance of about 5 mm from the free end of the workpiece. This choice was made after some preliminary tests to evalu-ate both the boundary effects on heat transfer between metal and environment and the effects of cutting fluid film on temperature distribution.

4   SURFACE EXAMINATION

After cutting tests, micrographic analyses were made using optical and SEM microscopes. The workpieces were cross sectioned (at 5 mm from the free end) and then polished and etched with a HNO_3 (1.0% vol)- HF (2% vol) solution.

The target of these examinations was to establish suitable cutting parameter values corresponding to lower microstructural alterations, in order to use them for mechanical components subjected to cyclic loads and specimens for fatigue tests.

As shown in Figures 3-5, machining causes little microstructural alterations; the pictures are referred to specimen 58A subjected to the heaviest cutting conditions (maximum cutting speed and maximum depth of cut). Figure 3 shows the material structure far from the cutting zone and Figures 4, 5 show the structure close to the generated surface; there are no substantial differences between them. Figure 6 shows a longitudinal section of the same specimen; it confirms the aforementioned observation.

Figure 7 shows the specimen 69B machined with minimum depth of cut, minimum feed and intermediate value of cutting speed. There is a remarkable grain distortion close to the generated surface. It is due to the relatively low value of titanium elastic modulus (about 110 GPa), so the workpiece has the tendency to move away from the cutting tool when a small depth of cut is used and the cutting operation will result in a crushing of the material.

Specimen 56B shows quite similar behaviour (Fig. 8) but the phenomenon is less evident than specimen 69B (depth of cut for 56B was 0.5 mm).

We can say that the higher the depth of cut the better the machinability of CermeTi-C-10 (this is well known for titanium alloys).

There are no difficulties in turning CermeTi-C-10; it needs only a little care about the choice of suitable cutting speeds because the presence of carbides in the composite requires the use of the lowest values within the recommended range for Ti6A14V (usually 6+46 m/min) (Donachie. Jr 1989).

Porosity is clearly shown in Figure 9 (and in Figure 3); the voids can reach average dimensions of about 10 microns. They are randomly distributed inside the bulk material so there is a high probability they will appear on the surface after cutting (Figure 10).

5   CONCLUSIONS

All the above items suggest the following conclusions.

In order to make cylindrical specimens for CermeTi-C-10 fatigue characterisation, a suitable surface finish is too difficult to obtain using enhanced machining operations. Crack initiation starts close to the tips of voids

27

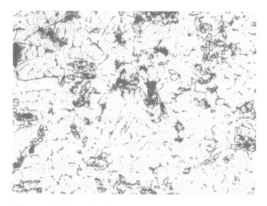

Figure 3. Test n° 58A: p=1.5; f=0.065; V_t=19.25. Microstructure far from cutting zone (250X).

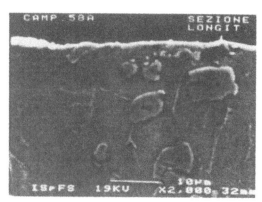

Figure 6. Test n° 58A: p=1.5; f=0.065; V_t=19.25. Microstructure near cutting zone: longit. section (SEM).

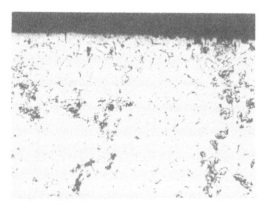

Figure 4. Test n° 58A: p=1.5; f=0.065; V_t=19.25. Microstructure near cutting zone: cross section (250X).

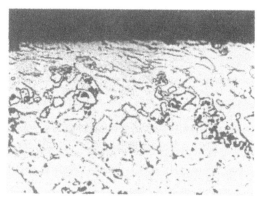

Figure 7. Test n° 69B: p=0.075; f=0.067; V_t=8.545. Microstructure near cutting zone: cross section (500X).

Figure 5. Test n° 58A: p=1.5; f=0.065; V_t=19.25. Microstructure near cutting zone: cross section (500X).

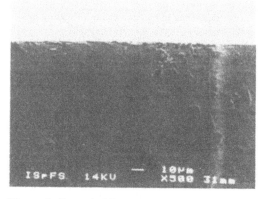

Figure 8. Test n° 56B: p=0.5; f=0.13; V_t=17.449. Microstructure near cutting zone: cross section (SEM).

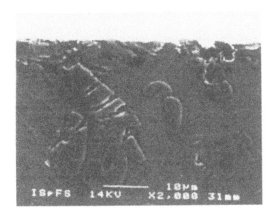

Figure 9. Test n° 56B: p=0.5; f=0.13; V_t=17.449. Carbides porosity will appear on the surface after cutting.

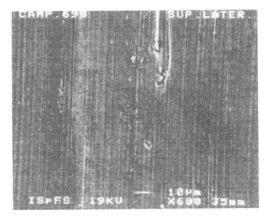

Figure 10. Carbides porosity appear on lateral surface after cutting.

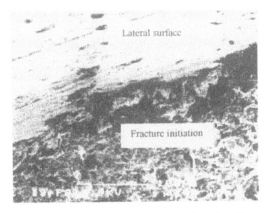

Figure 11. Fatigue fracture initiation on lateral surface.

appearing on the surface as shown in Figure 11 (referred to the fatigue fracture surface of a CermeTi-C-10 specimen), so it is often useless to take care about machining procedure to make specimens. This is true for CermeTi-C-10 but probably it is also true for similar materials.

Since the production of a higher density composite is more expensive, there are very few possibilities to obtain surfaces without flaws. This allows us to assert that, certainly for CermeTi-C-10, more sophisticated and expensive machining technologies to obtain better quality surface for specimens to use in fatigue characterisation, do not have real benefits. Therefore, standard procedure (ASTM STP 566) could be ignored and the optimisation of cutting parameters, at least for turning, (De Iorio et al. 1998), should not be of paramount concern.

The above suggestions point out the possibility of using technologies and procedures faster and cheaper because P M materials, such as CermeTi-C-10, have a mechanical behaviour that has low sensitivity with respect to surface finish.

6 ACKNOWLEDGEMENTS

The authors thank the laboratory technician of "Istituto Sperimentale Ferrovie dello Stato", Dr. Giuseppe Centolani, for the precious contribution he offered during this work.

REFERENCES

Harry E. Chandler 1989. *Machining of Powder Metallurgy Materals*. Metal Handbook-ASM International, 9th ed., Vol 16.

Matthew J. Donachie, Jr 1989. *Titanium. A technical Guide*. ASM International.

Cascetta F., Vigo P. 1988. *Le misure di Temperatura*. Liguori (Napoli-Italy) 2nd ed.

ASTM STP 566 1974. *Handbook of Fatigue Testing*.

A. De Iorio, D. Ianniello, R. Iannuzzi, F. Penta 1998. *Practice for Conducting Fatigue Tests at Various Temperatures for a MMC materials*. Proc. ECCM8, Naples (Italy), 3-6 June 1998.

Experimental Techniques and Design In Composite Materials 4, Found (Ed.)
© 2002 Swets & Zeitlinger, Lisse, ISBN 90 5809 370 0

Experimental determination of mechanical properties of polypropylene foams Neopolen P

I. Beverte
Institute of Polymer Mechanics, Riga, Latvia

ABSTRACT: Neopolen® P is a two-phase cellular composite material which for most practical applications is working in conditions of small as well as large compressive deformations. Therefore it is necessary to determine Young's moduli in compression and the compressive strength for a wide range of densities: $22.7 \text{ kg/m}^3 \leq \rho \leq 83.8 \text{ kg/m}^3$. Several experimental methods for the investigation of the mechanical properties of Neopolen are compared. The direct measurements on a universal testing machine are used for the main part of the experiments. The revealed anisotropy of the mechanical properties is correlated to the structure of the foams.

1 APPLICATIONS OF NEOPOLEN P

Neopolen® P (further referred to as Neopolen) is a two-phase cellular composite material consisting of a thermoplastic polymeric matrix (polypropylene) and a mobile gaseous phase. The structure of Neopolen is built up from pre-expanded polypropylene particles. On formation, automaton particles are pressed together under steam heat into details of different shape. The density of these details can be varied by choice of the degree of expansion of the particles. Neopolen products are characterised by their low weight, extremely good cushioning properties, and high energy absorption connected with good capability to retain the initial shape under dynamic loading. Additional features include shape stability over a wide temperature range ($-40°C \leq T \leq +110°C$), high stability against the influence of chemicals (solvents, oils, gases), low absorption of moisture and low heat transfer. These properties determine the range of applications of Neopolen products: 1) in machine building as an impact energy absorber and for the inside furniture of cars, 2) as packaging material for transportation of highly fragile products such as electronic devices, printers, personal computers, medical science products, 3) in house building as thermal insulators and construction elements, 4) in sport and entertainment industries as material for cushioning carpets. As a thermoplastic Neopolen can be readily recycled.

These excellent properties in a wide range of applications determine the necessity to investigate the structural mechanical properties of Neopolen in order to be able to manufacture Neopolen materials with a pre-assigned set of properties.

2 STRUCTURAL INVESTIGATIONS

The structure of Neopolen was investigated using both a light and raster-electronmicroscope (REM). These investigations have shown the characteristic two level structure of Neopolen. The first level is formed by pre-expanded polypropylene particles (Figure 1) pressed together with the help of steam. Figure 1 shows that a definite proportion of particles (which differ for each recipe) have a thin outer layer of monolithic polymer. The second structural level is formed by practically 100% closed cell isotropic foams inside the particles (Figure 2).

The results of dimensional measurements of the particles taken from photographs are presented in Table 1. The shape of the particles varies between that of cushion-like and an ellipsoidal cylinder. For all the series of particles investigated one dimension (l_z) of a particle is considerably smaller than the other two (l_x and l_y) ie $l_z < l_x$, l_y (where l_x, l_y, l_z are the average dimensions obtained from the measurement of ten particles). This leads to the existence of energetically favourable positions (positions of stable equilibrium) of a particle due to gravitational effects while pouring particles into the mould. Particles will tend to fall with the smallest dimension l_z parallel to the gravitational vector, the larger two being in perpendicular directions. With some allowance, particles of recipes N-8210 and N-9210 can be considered as having equal dimensions. Such particles will provide a random spatial distribution.

3 DENSITY DETERMINATION

Neopolen was received from BASF in the shape of plates with dimensions $L_x = 80\text{cm}$, $L_y = 51\text{cm}$,

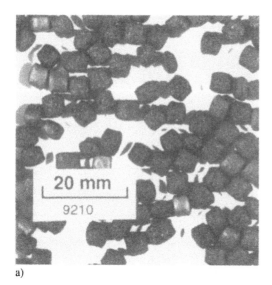 

a)   b)

Figure 1. Structural units of Neopolen: pre-expanded polypropylene particles before being pressed together. a) Neopolen-9210, b) Neopolen-9240 BSW (microphotographs acquired with the help of light microscope).

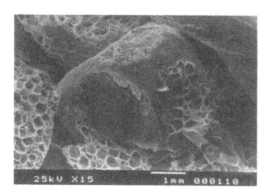

Figure 2. Break surface of Neopolen-9210 (a microphotograph acquired with the help of REM).

L_z = 15cm. It should be mentioned that in the present paper, as well as in other literature considering plastic foams, the term "density" is actually denoting the specific weight of foams (Берлин & Шутов 1980). The density ρ of each plate was determined by following the procedure in DIN 53420. For the compressive strength

samples measuring 50 x 50 x 50mm were used. While cutting out the samples on a bandsaw the more dense outer crust (which appeared during moulding) of each plate was cut off. The speed of the bandsaw should be chosen such that the polypropylene does not melt during cutting and no rigid outer crust appears on the samples.

The results obtained on two sets of samples are presented in Table 2. The density ρ was calculated together with the standard deviation s and the coefficient of variation v. The first set contained five samples for each density and the second set ten. The time difference of samples being cut and their density determination between both sets was 25 days in order to evaluate the possible influence of drying. The density of the Neopolen base polymer and polypropylene being the same ρ_0 = 902 kg/m^3 (Gibson & Ashby 1988), it is possible to determine the space filling coefficient Pl = ρ/ρ_0 of each plate.

It can be concluded that the materials investigated are highly homogeneous (coefficient of variation v ≤ 2%). In the limits determined by the standard deviation s the results obtained on both sets are the same. The space filling coefficient is: 2.5% ≤ Pl ≤ 9.3% so that the foams investigated do belong to the light ones

Table 1. Geometry of Polypropylene Particles

N	Neopolen recipe	Average dimensions of particles			Shape of particles
		l_x mm	l_y mm	l_z mm	
1	N-8210	3.8	4.3	2.9	Cushion - like
2	N-9210	4.8	4.1	2.9	Cushion - like
3	N-9225	3.4	4.1	2.0	Cushion - like
4	N-9230	2.9	4.4	1.6	Ellipsoidal cylinders
5	N-9240 BSW	2.6	3.9	1.6	Ellipsoidal cylinders

Table 2. Densities of the investigated Neopolen P plates

Plate N	Neopolen recipe	Manufacturing data	Data of density determination: 30.10.97		Data of density determination: 10.11.97	
			Density ρ kg/m^3	P1	Density ρ kg/m^3	P1
1	Neopolen P 9210	18.09.97	22.7 ± 0.2 * (± 0.9%)	2.5	-	-
2	Neopolen P 9210	18.09.97	22.7 ± 0.2 (± 0.9%)	2.5	(23.0 ± 0.5) (± 2.2%)	2.5
3	Neopolen P 9225	17.07.97	38.6 ± 0.3 (± 0.8%)	4.3	-	-
4	Neopolen P 9225	17.07.97	39.0 ± 0.3 (± 0.8%)	4.3	40.2 ± 0.7 (± 1.7%)	4.5
5	Neopolen P 9230	25.07.97	56.7 ± 0.7 (± 1.2%)	6.3	57.3 ± 1.2 (± 2.1%)	6.3
6	Neopolen P 9230	03.09.97	65.3 ± 0.3 (± 0.5%)	7.2	64.4 ± 0.6 (± 0.9%)	7.1
7	Neopolen P 9240-BSW	11.08.97	81.9 ± 0.4 (± 0.5%)	9.1	80.4 ± 1.1 (± 1.4%)	8.9
8	Neopolen P 9240 BSW	11.08.97	83.8 ± 1.0 (± 1.2%)	9.3	83.7 ± 1.7 (± 2.0%)	9.3

*) Here and later the second number is the standard deviation s, in parentheses - the variation coefficient v.

(Берлин & Шутов 1980). There is no pronounced trend in dependence of the density on the density determination time. It can be concluded that the Neopolen plates received were well dried.

4 YOUNG'S MODULI IN COMPRESSION

In most practical applications Neopolen works in conditions of small as well as large compressive deformations. Therefore it is necessary to investigate the deformation properties in compression. Young's moduli E_x, E_y, E_z were determined in compression according to DIN 53421. The most convenient method for determining Young's is by using a universal testing machine. A Zwick 100kN machine was used for the Neopolen compression tests. A computer was connected to the digital port to allow all the test results to be stored which were then processed using a commercial program ZDB. The program can present plots of σ-ε graphs to different scales, calculate Young's modulus for different regions of the σ-ε curve, simulate experiments under different initial conditions and carry out statistical processing of results.

At the same time it is important to estimate how precise are the values of moduli determined on the testing machine. Therefore measurements were made simultaneously according to the testing machine traverse displacement (with the help of a computer program ZDB, curve σ-ε) and with the help of the Boldwin-Hottinger extensiometer W5K (curve P-Δl on Yew Model plotter). Samples of dimensions 50 x 50 x 100 mm were placed with the biggest dimension 100mm parallel to the loading direction so that the inductive extensiometer W5K could be attached in the gap between the loading surfaces of the machine platens (see Figure 3).

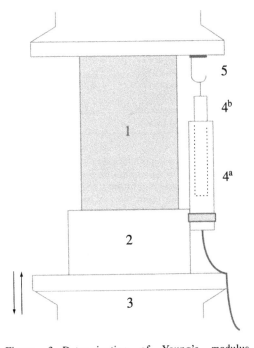

Figure 3. Determination of Young's modulus simultaneously on Universal testing machine and Inductive extensiometer: 1 - Neopolen sample 100 x 50 x 50 mm; 2 - a steel pad; 3 - testing machine's traverse; 4 - inductive extensiometer W5K: 4[a] - a coil, 4[b] - a ferromagnetic rod; 5 - a hook glued to the upper top of testing machine.

Table 3. Measurements of Young's modulus E_z

N	Density ρ	Testing machine traverse		Inductive extensiometer W5K	
		Modulus E_z MPa	Average modulus E_z MPa	Modulus E_z MPa	Average modulus E_z MPa
1	22.9 kg/m^3	3.00 3.00 2.86	2.99 ± 0.12 (± 4.0%)	3.06 3.07 2.70	2.94 ± 0.21 (± 7.2%)
2	84.1	3.00 27.7 30.7	29.47 ± 1.57 (± 5.3%)	29.8 27.8 30.6	29.4 ± 1.44 (± 4.9%)

Table 4. Young's Moduli

Plate N	Neopolen recipe	Density ρ kg/m^3	Modulus E_x MPa	Modulus E_y MPa	Modulus E_z MPa
1	Neopolen P 9210	22.7	(2.8 ± 0.1) (± 3.6%)	(2.8 ± 0.0) (± 0.0%)	(2.8 ± 0.1) (± 3.6%)
2	Neopolen P 9210	22.4	2.2 ± 0.2 (± 9.1%)	2.3 ± 0.2 (± 8.7%)	2.3 ± 0.0 (± 0.0%)
3	Neopolen P 9225	38.6	8.0 ± 0.6 (± 7.0%)	8.5 ± 0.2 (± 2.3%)	9.2 ± 0.4 (± 4.3%)
4	Neopolen P 9225	39.0	7.8 ± 0.2 (± 2.3%)	8.5 ± 0.2 (± 2.3%)	9.2 ± 0.1 (± 1.1%)
5	Neopolen P 9230	56.7	14.2 ± 0.2 (± 1.4%)	15.1 ± 1.0 (± 6.6%)	17.0 ± 0.5 (± 2.9%)
6	Neopolen P 9230	65.3	20.3 ± 0.8 (± 3.9%)	16.0 ± 0.9 (± 5.6%)	17.3 ± 1.7 (± 9.8%)
7	Neopolen P 9240-BSW	81.9	27.2 ± 1.9 (± 7.0%)	22.0 ± 0.5 (± 2.3%)	22.8 ± 0.3 (± 1.3%)
8	Neopolen P 9240-BSW	83.8	26.4 ± 2.8 (± 11.0%)	22.3 ± 2.2 (± 9.9%)	23.4 ± 1.0 (± 4.3%)
8	Neopolen P 9240-BSW	84.77	29.0 ± 1.0 (± 3.4%)	24.7 ± 3.1 (± 13.8%)	25.3 ± 1.5 (± 5.9%)

The ratio of the sample's height to width of 100 mm/ 50 mm = 2 ensures the elimination of end zone influence on measurements (Hilyard 1982). An inductive extensiometer was used for precise measurements because it was practically impossible to glue strain gauges on to the polypropylene. The stiffness of a strain gauge itself can influence the measurement of deformation properties of Neopolen having low Young's modulus: E < 30 MPa. In addition while measuring, strain gauges can warm up considerably and change the deformation properties of plastics. The experiment was carried out for two recipes: the lightest one N-9210 (ρ = 22.9 kg/m^3) and the heaviest one N-9240-BSW (ρ = 84.1 kg/m^3) with a loading rate ε = 10%/min. The results obtained are presented in Table 3. Calculations are carried out with increased accuracy (two decimal characters) in order to emphasise the investigated effect. The results permit to draw the following conclusions.

1. For Neopolen plastic foams, measurement of the modulus E_z according to the testing machine traverse and to the inductive extensiometer are providing practically the same results. This can be explained by the low stiffness of the foams investigated: 3 MPa ≥ E_z ≥ 30 MPa. For such materials small loading forces and large displacements are characteristic:

N9210: when P_H >>8>>1.25 mm;

N-9240-BSW: when P_H >>780 N, $\Delta l z$ >>0.85mm; where P_H, $l z_H$ are load and displacement at the end of Hooke's region. Under such conditions the stiffness of the testing machine itself has negligible influence on the measurements according to the traverse displacement. (The same experiment performed on the same testing machine for a metal sample provided a 2 to 3 times difference in modulus values.)

2. As the values of moduli E_x, E_y are of the same order as that of E_z the same statements can be made considering measuring of these moduli on the testing machine.

3. It can be concluded that the testing machine can be used for sufficiently precise determination of the Young's moduli of Neopolen in compression.

Following the results in Table 3 the main series of Young's moduli determinations was performed on the testing machine. To investigate the end zone influence samples had dimensions 50 x 50 x 50 mm. Samples were loaded in three mutually perpendicular directions. The linear Hooke's region lays in the following limits for all investigated densities:

$$0.2\% < \varepsilon_{xx}^{lin}, \varepsilon_{yy}^{lin}, \varepsilon_{zz}^{lin} < 1.3$$

Each value of modulus E was calculated from three separate measurements. The results obtained are presented in Table 4. Several conclusions can be made. In the limits of standard deviation the lightest foams (Neopolen-9210) are practically isotropic:

$$\rho = 22.74 \text{ kg/m}^3; E_x \sim E_y \sim E_z$$

Foams with intermediate densities (Neopolen-9225, the lightest plate of Neopolen-9230) have a monotropic axis parallel to the oz dimension of the moulding plate:

$$38.6 \text{ kg/m}^3 \leq \rho \leq 56.7 \text{ kg/m}^3; E_z > E_x, E_y; E_x \sim E_y.$$

The heaviest ones (Neopolen-9230, Neopolen-9240-BSW) are well expressed monotropically too. The axis of monotropy is parallel to the ox dimension of the plate:

$$65.3 \text{ kg/m}^3 \leq \rho \leq 84.77 \text{ kg/m}^3; E_x > E_y, E_z; E_y \sim E_z.$$

No Neopolen material turned out to be orthotropic. These results can be explained in the following way. The determined mode of anisotropy for each recipe is a result of complex interaction of two phenomena:

1) the tendency of polypropylene particles with one dimension exceeding the other two to assume positions of stable equilibrium under action of the earth's gravitation vector g.

2) electrostatic charges between particles in the process of pouring them into the mould.

The first phenomenon leads to a certain regularity in the spatial distribution of elongated particles, ie to the existence of a plane of isotropy and an axis of monotropy. The monolithic, more dense outer crust of elongated particles provide different amounts of base polymer in different directions. This leads to monotropy of the resulting material. It is dependent on the degree of elongation of $l_{zx} = l_z/l_x$, $l_{zy} = l_z/l_y$ and the weight P_i of a particle.

The electrostatic charges being independent of the earth's gravitation direction tends to cause random distribution of the particles. The strength of the electrostatic forces arising from the mutual friction between particles is dependent on the area of the outer surface (ie volume) of particles. For particles pre-expanded to larger volume (N-9210, N-9225) the electrostatic forces exceed the tendency of particles to assume the energetically favourable positions. For particles having smaller volume and small parameters l_{zx}, l_{zy} (N9225, N930, N99240BSW) the electrostatic forces are not strong enough to destroy the distribution leading to monotropy. To apply these considerations to a definite material it is necessary to know the direction of the earth's gravitation vector g in respect to the mould dimensions while pouring the particles. In order to

understand why there are two different axes of monotropy additional investigations are necessary.

The dependence of E_x, E_y and E_z on foam density has practically a linear characteristic as shown in Figure 4. As the Young's modulus of rigid polypropylene $E_0 = 1130$ MPa (Gibson & Ashby 1988), it is possible to calculate the relative Young's moduli E_x/E_0, E_y/E_0, E_z/E_0 of Neopolen.

Results of modulus E_z obtained on samples with dimensions 50 x 50 x 100 mm are approximately 20% higher than those obtained on cubic samples 50 x 50 x 50 mm. This corresponds well with the results of other authors (Hilyard 1982) and can be explained with elimination of the above mentioned end zone effect. The results of modulus E determined on 100 x 50 x 50 mm samples are considered as more precise because the central zone of highly homogeneous stress-strain state is relatively larger. When loading on cubic samples (in order to spare material or when having thin plates) this is a symmetric error that should be taken into account.

5 COMPRESSION STRENGTH

The main practical applications of Neopolen lie in the field of large deformations (above Hooke's region). Therefore it is necessary to know the deformation and strength properties of Neopolen under large deformations. Typical stress-strain curves σ_{xx}-ε_{yy} are presented in Figure 5 (σ_{yy}-ε_{yy} and σ_{zz}-ε_{zz} differ very slightly). For all examined densities the σ-ε curves do show a well expressed second linear region when $5\% < \varepsilon < 40\%$. So it is possible to determine modulus E2 in this linear region corresponding to gradual buckling of structural elements. Practically, E2 was calculated for $12.5\% < \varepsilon < 37.5\%$. The results obtained are presented in Table 5.

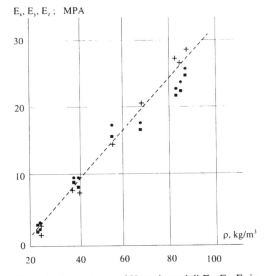

Figure 4. Dependence of Young's moduli E_x, E_y, E_z in compression on foams density. $+$ - E_x, $\blacksquare$ - E_y, $\bullet$ - E_z.

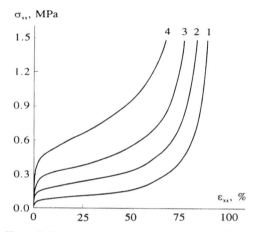

σ$_{xx}$, MPa

Figure 5. Stress-strain curves in compression parallel to ox direction. 1) ρ = 23.0 kg/m^3; 2) ρ = 40.2 kg/m^3; 3) ρ = 57.3 kg/m^3; 4) ρ = 83.7 kg/m^3.

Modulus E2 is approximately 10-30 times smaller than E. Some monotropy can be observed but of a different mode than that of E.

There are no definite criteria for the determination of the compression strength of Neopolen. In the present investigation the compression strength is determined at that point of the σ-ε diagram where the stress starts to increase rapidly (the σ-ε curve goes parallel to the σ axis). The samples were sufficiently elastic to retain structural unity even at 90% deformations. In Figure 6 the stress in dependence of density (see also Table 6) is presented for strains of 25% < ε < 87.5%.

No expressed dependence of stress on the sample's cutting direction can be observed. The material is slightly orthotropic.

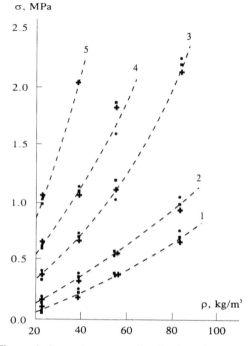

σ, MPa

Figure 6. Stress in compression in dependence of density. + - σ$_{xx}$, ■ - σ$_{yy}$, ● - σ$_{zz}$. 1) ε$_{xx}$, ε$_{yy}$, ε$_{zz}$ = 25 %, 2) 50%, 3) 75%, 4) 81.25%, 5) 87.50%.

6 AVERAGE LIMIT STRESS IN BENDING

A 3-point bend experiment was performed according to DIN 53423. Samples were cut out from the plate in six modes in order to examine the potential anisotropy

Table 5. Modulus E2 in second linear region

Plate N	Density ρ kg/m^3	Modulus E2$_x$, MPa	Average modulus E2$_x$, MPa	Modulus E2$_y$, MPa	Average modulus E2$_y$, MPa	Modulus E2$_z$, MPa	Average modulus E2$_z$, Mpa
2	23.0	0.19 0.19 0.19	0.19 ± 0.00 (± 0%)	0.23 0.19 0.23	0.22 ± 0.02 (± 9%)	0.19 0.23 0.23	(0.22 ± 0.02) (± 9%)
4	40.2	0.32 0.32 0.32	0.32 ± 0.00 (± 0%)	0.32 0.32 0.32	0.32 ± 0.00 (± 0%)	0.36 0.36 0.36	0.36 ± 0.00 (± 0%)
5	57.3	0.52 0.39 0.45	0.45 ± 0.07 (± 14%)	0.58 0.55 0.55	0.56 ± 0.02 (± 3%)	0.52 0.51 0.55	0.53 ± 0.02 (± 5%)
8	83.7	0.89 0.91 0.89	0.90 ± 0.01 (± 1%)	0.91 0.91 0.91	0.91 ± 0.00 (± 0%)	1.00 0.91 0.96	0.96 ± 0.05 (± 5%)

Table 6. Compression strength

Plate N	Density ρ kg/m^3	Deformations ε_{xx}, ε_{yy}, ε_{zz} %	Stresses		
			σ_{xx} MPa	σ_{yy} MPa	σ_{zz} MPa
2	23.0	25.00	0.10	0.11	0.10
		50.00	0.16	0.16	0.17
		75.00	0.42	0.42	0.42
		81.25	0.61	0.61	0.60
		87.50	1.07	1.02	1.02
4	40.2	25.00	0.225	0.25	0.27
		50.00	0.33	0.36	0.37
		75.00	0.72	0.75	0.77
		81.25	1.08	1.09	1.16
		87.50	2.08	-	-
5	57.3	25.00	0.40	0.42	0.38
		50.00	0.55	0.60	0.54
		75.00	1.15	1.21	1.05
		81.25	1.83	1.84	1.60
		87.50	-	-	-
8	83.7	25.00	0.67	0.70	0.73
		50.00	0.93	0.95	1.05
		75.00	2.10	2.17	2.23
		81.25	-	-	-
		87.50	-	-	-

Table 7. Dimensions of samples for different cutting modes

Mode N	l_x mm	l_y mm	l_z mm
1	120 ± 1	25 ± 1	25 ± 1
2	120	25	25
3	25	120	20
4	25	120	20
5	20	20	120
6	20	20	120

(see Table 7). In two cases the longest dimension of a sample was parallel to the ox asix, in two to the oy axis and in two to the oz axis.

The testing machine with a load range of 5kN was used, the loading speed (the speed of the loading pin) was 1 mm/min. As the 3-point bend test is not representing pure bending this experiment cannot be used for the determination of the modulus of elasticity. None of the samples broke, having achieved a deflection of 20 mm. In this case according to the mentioned standard the limit stress σ_{20} at 20 mm bend should be determined as a characteristic of the bending properties of a material. The results presented in Table 8 permit the author to draw the following conclusions:

1) no dependence between the limit stress σ_{20} and the cutting mode of samples can be observed.

2) the dependence of σ_{20} on the density is linear for all six cutting modes.

Table 8. Limit stress σ_{20} in bend

Plate N	Density ρ kg/m^3	Average limit stress σ_{20} in bend MPa					
		Mode 1	Mode 2	Mode 3	Mode 4	Mode 5	Mode 6
2	23.0	0.027 ± 0.001 (± 3%)	0.027 ± 0.000 (± 2%)	0.027 ± 0.001 (± 4%)	0.027 ± 0.000 (± 2%)	0.030 ± 0.000 (± 1%)	0.030 ± 0.000 (± 2%)
4	40.2	0.059 ± 0.003 (± 5%)	0.058 ± 0.003 (± 6%)	0.057 ± 0.001 (± 1%)	0.057 ± 0.001 (± 2%)	0.058 ± 0.001 (± 2%)	0.058 ± 0.002 (± 4%)
5	57.3	0.115 ± 0.004 (± 3%)	0.120 ± 0.007 (± 6%)	0.115 ± 0.006 (± 6%)	0.118 ± 0.002 (± 2%)	0.117 ± 0.003 (± 3%)	0.110 ± 0.004 (± 4%)
8	83.7	0.176 ± 0.005 (± 3%)	0.168 ± 0.002 (± 1%)	0.195 ± 0.012 (± 6%)	0.182 ± 0.005 (± 3%)	0.180 ± 0.002 (± 1%)	0.196 ± 0.007 (± 4%)

The structural and mechanical properties of Neopolen are investigated in a wide range of densities. The main result of the investigation is the following: in the region of linearity the stress-strain curves of Neopolen P can be measured on a Zwick (100 kN) universal testing machine with a high degree of precision. Time and money are saved because there is no need to carry out simultaneous expensive precise measurements (using strain gauges, adhesives and inductive extensiometers). Therefore the main series of Young's moduli determinations was performed on the testing machine.

In preliminary technical information (Vorlaufige techn. Inform. 1992) for Neopolen the manufacturers (BASF Ltd) considered it as an isotropic material. The present investigation has proved that Neopolen posses well expressed anisotropy. The lightest Neopolen P recipes turned out to be isotropic, the medium and heaviest ones monotropic. At large deformations (compression as well as bending) Neopolen exhibits no definite anisotropy.

Structural investigations are correlated with mechanical tests. By varying the degree of the particles' elongation as well as the degree of pre-expansion of the particles, Neopolen materials with pre-assigned properties can be projected and obtained. This is extremely important for practical applications.

REFERENCES

А.А.Берлин, Ф.А.Шутов. Химия и технология газонаполненных высокополимеров, М., 1980. (In Russian.)

Lorna J. Gibson, Michael F. Ashby. Cellular Solids. Structure and Properties. Pergamon Press, first edition 1988.

Mechanics of Cellular Plastics. Edited by Hilyard N.C. London, 1982.

Neopolen P. Vorläufige technische Information, BASF, HSB/S, N4, 1992. (In German.)

Experimental Techniques and Design In Composite Materials 4, Found (Ed.)
© 2002 Swets & Zeitlinger, Lisse, ISBN 90 5809 370 0

Test methods for composites mechanical characterisation

A. De Iorio, D. Ianniello, R. Iannuzzi & F. Penta
Dipartimento di Progettazionne e Gestione Industriale, Università Federico II, Napoli

A. Apicella & F. Quadrini
Dipartimento di Ingegneria dei Materiali e della Produzione, Università Federico II, Napoli

ABSTRACT: The goal of every mechanical engineering activity is to design and produce components that principally must operate in safe conditions. Hence the knowledge of the material behaviour and strength is an essential key. The latter aspect become more complex in dealing with metal matrix composite materials than traditional ones, owing to the wider variety of the failure modes and the consequent difficulty in modelling their overall mechanical behaviour. In the present paper the Authors review the most important test methodologies employed to characterize mechanical behaviour of composite materials: particular regard is given to the specimen preparation and to the related theoretical aspects that are essential in order to have cautious use of the results.

1 INTRODUCTION

Composite characterisation tests are the basis for modelling the thermo-mechanical behaviour of these materials. Furthermore, they permit to evaluate characteristic parameters necessary for the structural analysis. For unidirectional composites Table. 1.1 shows a list of the mechanical properties that arc fundamental for the application of the most important structural theories. In order to obtain these values, composites can be tested as the traditional isotropic materials are, even if anisotropy, brittleness of the reinforced phase and lower strength of the matrix have to be taken into account. Moreover mechanical properties evaluated with standard specimens could lead to erroneous predictions of the behaviour of complex components. In fact, volumetric ratio of the reinforcement, its distribution, polymerisation or sintering conditions could vary along the geometry. In this case, a specimen with the same geometry as the real component is useful to estimate the mechanical behaviour of the latter.

In this study an outline of tests is described that could permit the evaluation of the mechanical properties listed in Table 1.1 for metal matrix composites, which are commonly available as laminae or plane laminates. Reference standards for these tests are mentioned, when available.

2 TENSILE TEST IN PRINCIPAL DIRECTIONS

K. G. Kreider (1969) observed that the ASTM-E8 standard established specimens with shape and dimensions inadequate for characterising boron-aluminium composites. In fact in tensile tests, premature failure occurred in the gripping areas, because the shear strength of the matrix did not permit the transfer of the high tensile loads, which produced specimen failure in the useful length. Kreider described two ways the cracks occurred. The first affected the material layers near the grips and was displaced along the whole width W_G (Fig. 2.1).

Table 1.1. Mechanical properties of unidirectional composites.

Property	Symbol
Longitudinal Young's modulus	E_x
Transverse Young's modulus	E_y
Poisson's ratio	ν_{xy}
Shear modulus in the plane of the lamina	G_{xy}
Longitudinal tensile or compressive strength	σ_l^+, σ_l^-
Transverse tensile or compressive strength	σ_t^+, σ_t^-
Shear strength	τ_S
Fracture toughness	K_{Ic}, K_{IIc}
Fatigue curve	

The second was distributed longitudinally in the line that separated the shoulders from the central section whose width is W_R. In order to avoid the former kind of cracks, the gripping area needs to be opportunely extended. If P_{max} is the tensile force that generates failure in the useful length of the specimen, it can be considered:

$$P_{\max} = \sigma_R t W_R \qquad (1)$$

where t is the specimen thickness and σ_R is the rupture stress in the test direction. The gripping area $W_G l_G$, which permits the maximum load transfer, can be evaluated by the following equation:

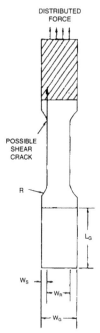

DISTRIBUTED
FORCE

POSSIBLE
SHEAR
CRACK

R

l_G

W_S

W_R

W_G

Figure 2.1. Tensile test specimen.

$$2(W_G l_G)\tau_G = \sigma_R t W_R \qquad (2)$$

where l_G is the length of the contact area and τ_G is the mean value of shear stress in the axial direction. Assuming τ_G is equal to the matrix shear strength τ_C, the minimum gripping area is:

$$W_G l_G = \frac{t W_R}{2} \frac{\sigma_R}{\tau_C}. \qquad (3)$$

Because of the high ratio σ_R/τ_C high gripping areas are required. In order to avoid the second kind of cracks, it is necessary to reduce the mean shear stress τ_m which acts along the line which separates the shoulders from the central zone. In an approximate way τ_m can be evaluated by supposing that the axial force P is uniformly distributed on the contact areas and the load, which acts on the shoulder, is transferred by shear to the gripping central zone. From a force balance on a shoulder:

$$\tau_m l_G t = \frac{P}{2W_G l_G}(2W_S l_G). \qquad (4)$$

In the condition of incipient specimen failure, the load P and the mean shear stress τ_m reach the respective maximum values P_{max} and τ_m^{max}; we have:

$$\tau_m^{max} l_G t = P_{max} \frac{W_S}{W_G} = \sigma_R t W_R \frac{W_S}{W_G} \qquad (5)$$

and then:

$$\tau_m^{max} = \frac{\sigma_R W_R}{W_G l_G} W_S = \frac{\sigma_R W_R}{W_G l_G} \frac{(W_G - W_R)}{2}. \qquad (6)$$

Low values of Ws and high values of l_G yield to low τ_m^{max}. If Ws=0 the mean shear stress τ_m becomes zero: in this case the stress concentration effect on the shoulders is also eliminated and the specimen can be realised with easier processing. But the stress concentration in the contact areas, which transfer loads and alignment errors of test machines, cause additional stresses which could generate failure in the non useful zones near the grips.

Stress concentration is reduced by placing strengthening tabs of ductile material upon the load transfer areas. These tabs are opportunely tapered towards the central part of the specimen. Tabs also protect specimens from the contact action of the grips. Alignment errors can be reduced by using self-aligning devices.

Kreider (1969) introduced equations (3) and (6). These derive from an approximate approach, which neglects the real complex stress distribution in the specimen. Equations are useful because they show the geometric parameters which influence the premature failure and which have to be altered for a correct specimen sizing. Generally specimen choice requires an exact analysis of the stress distribution which depends on the same material constant being determined by testing. Thus a trial and error procedure is necessary, changing at every step the sizes which seem to influence the premature failure.

During a test, strain can be measured using strain gauges or extensometers. In both cases high measuring sensitivity is required because the strain to failure is very low (about 0.01) and the variations in slope of the experimental curve are small when changes in mechanical behaviour occur. To evaluate the presence of possible bending strains, longitudinal strains have to be measured on both faces of the useful part of the specimen. The reference length for strain estimation has to include a statistically significant area of the material. Using extensometers has problems related to the high surface hardness, due to the reinforced phase, and to the necessity of applying low forces between the instrument rods and the specimen.

The combination of these two factors often leads to the slipping of the rods over the specimen.

For the described test there are several standards: among these ASTM is the most used. It provides recommendations for specimen geometry (ASTM D 3552), specimen machining (ASTM E 8), verification of testing machine (ASTM 4) and verification and classification of extensometers (ASTM 83).

3 COMPRESSIVE TEST IN PRINCIPAL
  DIRECTIONS

Uniaxial compressive tests in the material principal directions present the same problems as the tensile tests and also buckling phenomena. For this reason specimens are like those for tensile tests with the addition of suitable constraints that avoid displacements normal to their axis. Generally specimens (Fig. 3.1) are flat, without shoulders and tabbed with adequately long and tapered steel or glass/epoxy tabs which limit buckling

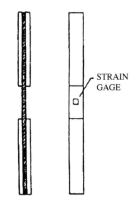

Figure 3.1. Compressive test specimen.

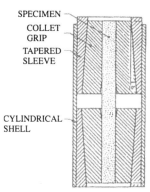

Figure 3.2. Celanese test fixture.

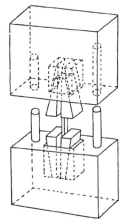

Figure 3.3. IITRI test fixture.

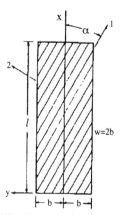

Figure 4.1. Off-axis test specimen geometry.

problems. These specimens can be used in Celanese or IITRI tests (ASTM 1987). In the former (Fig. 3.2) the axial load is transmitted by means of two conical collect grips in which two tapered sleeves are placed. The sleeves clamp the specimen. All the system is positioned into a cylindrical shell which forces the sleeves to move axially in order to avoid the additional bending stress. This device is disadvantageous because it requires the perfect coupling of the conical surfaces, even though the tabs have non-uniform thickness.

The IITRI test (Fig. 3.3) eliminates this factor using two pairs of trapezoidal wedges for specimen clamping.

Both the Celanese and IITRI tests are standardised from the ASTM for the characterisation of long fiber and polymeric matrix composites.

4   OFF-AXIS TENSILE TEST

The off-axis tensile test is frequently used in order to study elastic and strength properties of anisotropic fiber reinforced materials. The test consists of applying a tensile load on a flat rectangular coupon. Coupons are obtained from laminae oriented at an angle $\alpha \neq 0$

between the reinforced fiber direction and applied load direction (Fig. 4.1). In the off-axis tensile test, the elastic strains are measured and the applied load is registered, so that the normal elastic modulus E_x in the load direction is evaluated. Afterwards the elastic shear modulus G_{12}, referred to the principal material directions of the lamina plane, can be calculated. Furthemore, several material constants present in the most use failure conditions can be achieved from the tensile strength $\overline{\sigma_x}$ at different α values.

Considering only a uniform tensile stress field σ_x applied to the specimen, a longitudinal length variation $\Delta l = \dfrac{\sigma_x}{E_x} l$ and a transverse contraction $\Delta w = -\nu_{yx} \dfrac{\sigma_x}{E_x} w$ occur.

Together with these two deformations also a distortion from the rectangular shape to a parallelogram happens, by means of the shear strain $\gamma_{xy} = S_{16}\sigma_x$ (with S_{16} the 1-6 term of the elastic compliance matrix $[S_{ij}]$). Clamping devices prevent this distortion so that

41

additional stress fields are added to the uniform stress σ_x. Pagano & Halpin (1968) first noted this effect.

They obtained an analytical solution, of the elastic equilibrium equations of the specimen with the assumptions of plane stress state and only y dependence for the shear stress τ_{xy}. They also imposed the following boundary conditions:

$$u(0,0) = 0,$$

$$v(0,0) = \frac{\partial u}{\partial y}(0,0) = 0,$$

$$u(l,0) = \overline{\Delta l},$$

$$v(l,0) = \frac{\partial u}{\partial y}(l,0) = 0,$$

where u and v are the displacement components respectively in the x and y direction and $\overline{\Delta l}$ is the longitudinal length change imposed by the test machine. Solving elastic equilibrium problem also leads to identify the relation between the measured modulus E_x^* and the real one E_x:

$$E_x^* = E_x \frac{1}{1-\eta}$$

where η assumes the meaning of an error estimation for the modulus measured; its expression is:

$$\eta = \frac{6 S_{16}^2}{S_{11}\left(6 S_{66} + \dfrac{l^2}{h^2} S_{11}\right)}.$$

The magnitude of additional stresses can be limited using the fixture Wu & Thomas (1968) introduced (Fig. 4.2). The particular shape of the straightening tabs permits specimen constraint which allows free rotations to the boundary sections of its useful area. The same authors demonstrated the fixture's effectiveness. They measured strains at the points indicated in Fig. 4.2 by placing strain gauges on graphite/epoxy specimens with ratios $l/w = 2.5$, 4 and 5. In Fig. 4.3 the measured strain component values ε_x, ε_y and γ_{xy} are shown together with the corresponding values obtained from the analytical solution of Pagano & Halpin. All the curves are referred to the mean value p of the stress σ_x which acts in the useful part of the specimen.

If in the specimen only a uniform stress component σ_x would act, then we will have:

$$\frac{\varepsilon_x}{p} = 0.97,$$

$$\frac{\varepsilon_y}{p} = 0.317,$$

$$\frac{\gamma_{xy}}{p} = 2.34.$$

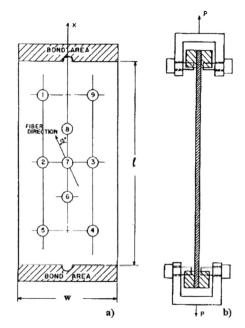

Figure 4.2. (a) Specimen with strain-gauge location; (b) Rotating clamp device.

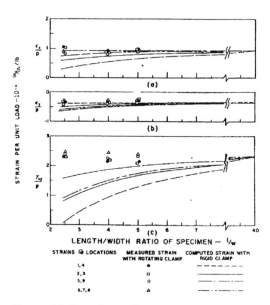

Figure 4.3. Experimentally measures strains with rotating clamp (Wu & Thomas, 1968).

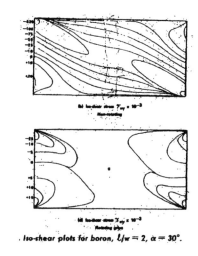

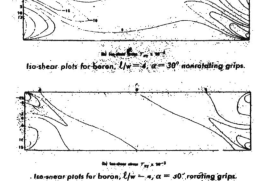

Figure 4.4. Iso-shear stress curves (Rizzo, 1969).

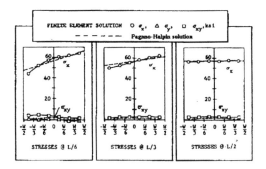

Figure 4.5. Stress distribution curves for a boron/epoxy specimen with $l/w = 6$ (Richards et al. 1969).

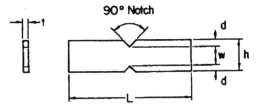

Figure 5.1. Iosipescu specimen.

These values and the experimental ones are similar so that it can be assumed additional stresses are negligible in the area delimited by the points 1, 5, 4 and 9. Anyway additional stresses are concentrated at the boundaries of the useful specimen part, where tabs prevent transverse contractions. Fig. 4.4 (Rizzo 1969) shows iso-shear stress curves for a boron/epoxy specimen. They are achieved by means of a finite element analysis both in the usual case of non-rotating grips and of the case of rotating grips proposed by Wu & Thomas. The latter gives good accuracy in the normal elastic modulus evaluation, but tensile strength estimation remains problematic.

In fact, failure could occur in the central zone where a uniform stress acts or in the external ones where the additional stresses are present. A specimens fails in one way or the other, depending on the specific material properties.

Richards et al. (1969) proposed the use of slender specimens with length-width ratio $l/w \sim 12$. In this case

because of the low specimen stiffness, clamping produces negligible additional stresses compared to the axial load. Furthermore, additional stresses decay before affecting the useful zone. These considerations are validated by a finite element analysis performed by the same authors (Fig. 4.5) assuming $l/w = 6$ and $\alpha = 45°$.

In order to avoid failure in the marginal zones of the useful part of the specimen, the edges can be strengthened with a further couple of tabs, realised in compliant material (like glass-epoxy-woven). In this way specimen stiffness essentially does not change and more applied load is transferred by the tabs until failure occurs in the useful central part. Nowadays this kind of test is not normally used.

5 IOSIPESCU SHEAR TEST

By means of the Iosipescu test (1967) a pure shear stress field is induced in the mid length cross section of a flat, thin specimen where two 90° V-notches are realised (Fig. 5.1). An Iosipescu specimen of an isotropic and homogeneous material, constrained as in Fig. 5.2, behaves like a plate which is physically and geometrically symmetric about the y axis and loaded by a two force, system with asymmetric resultants about the same axis. From the translation of equilibrium of half the specimen along the y direction, the following is obtained:

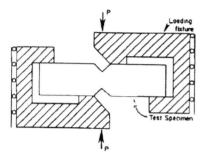

Figure 5.2. Loading fixture for Iosipescu test.

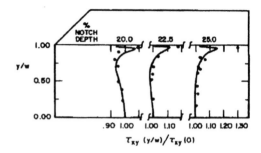

Figure 5.3. Experimental data (•) compared to numerical results (-) (Sullivan et al. 1984).

$$V = P \qquad (1)$$

where P is the load applied to the fixture by the testing machine and V is the resultant of the shear stress τ_{xy} which act in the middle cross section. The stress concentration effect due to the notches can produce a uniform shear stress area around the longitudinal specimen axis. An analytical expression of the stress field is not available for the notched part. By means of photoelastic analysis Iosipescu (1963) observed that the shear stress distribution in the reduced section depended on the d/h ratio.

Raising this ratio, the shear stress distribution previously presented a relative maximum after a minimum near the longitudinal specimen axis. According to Iosipescu's data, d/h values between 0.20 and 0.25 generate a shear stress field that can be approximately considered uniform on almost all of the middle cross section. Only near notches are there significant variations. So with this assumption for the d/h ratio, the shear stress is with good approximation given by:

$$\tau_{xy} = \overline{\tau}_{xy} = \frac{P}{t(h - 2d)} \qquad (2)$$

where $\overline{\tau}_{xy}$ is the mean shear stress value. Measuring the shear strains that occur in the central zone, equation (2) permits easily an experimental evaluation of the shear modulus G_{xy}.

Iosipescu's conclusions were validated later by Sullivan et al. (1984). They repeated the photoelastic measurements with more accuracy. In Fig. 5.3 their experimental values are reported together with numerical results of a finite element analysis they performed too. They confirmed that assuming $0.2 \le d/h \le 0.25$, the shear stress distribution is nearly uniform in approximately 85% of the reduced section extension. Here the deviation of the shear stress τ_{xy} from its mean value $\overline{\tau}_{xy}$ is always less than 3%.

The symmetrical arrangement of the specimen about the load action line strongly affects the test efficiency. Furthermore, loads have to be transmitted on the specimen through contact areas enough distance from the notches in order to avoid normal stress σ_y and shear stress τ_{xy} which have comparable values in the reduced section.

Sullivan et al. (1984) suggested a 10 mm distance between the inner loading points and the notch-root axis, meanwhile Adams & Walrath (1987b) proposed a 6.4 mm (0.25 in) distance. The lateral guides of the constraining device prevent a global rotation under the additional moment induced by a possible non-alignment of the machine. Guides also limited displacement amplitudes in the measurement area. High values of these displacements could invalidate the use of equation (2).

The suitability of the test for shear strength estimation seems strictly linked to the nature of the various materials. Adams & Walrath (1987a) observed that 6061-T651 aluminium alloy specimens failed as a consequence of the formation of a narrow, highly yielded area in the notched section. This ensured that the shear stress was uniform in that zone and so equation (2) was valid. For brittle materials, the suitability depends also the way failure occurs. In fact, Sullivan et al. (1984) noted that Derakene 470-36 (Dow Chemical) vinylester resin specimens failed due to the sudden propagation of cracks formed on the notch faces in the planes on which the maximum tensile stress acted. Crack initiation takes place by means of a uniaxial stress field, furthermore the photoelastic analysis showed that:

$$\sigma_1 = \overline{\tau}_{xy} \qquad (3)$$

where σ_1 is the only non-zero principal stress. The authors concluded that failure occurred owing to the cleavage under a tensile stress practically equal to the maximum value that the mean shear stress reached in testing. In fact the tensile strengths σ_1 evaluated in this way agreed with those evaluated with a tensile test.

The Iosipescu specimen can be used in another kind of test called the asymmetric four point bend test (AFPB test). Slepetz et al. (1978) first introduced this test with a device like that depicted in Fig. 5.4. The equilibrium equation for the reduced section is now:

$$V = P \frac{a - b}{a + b} \qquad (4)$$

and the mean shear stress is:

$$\overline{\tau}_{xy} = \frac{P}{t(h - 2d)} \frac{a - b}{a + b}. \qquad (5)$$

44

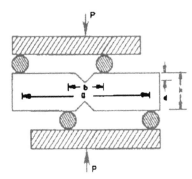

Figure 5.4. The asymmetric loading of an Iosipescu specimen.

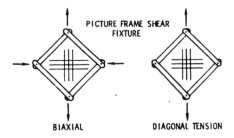

Figure 6.1. Shear test fixture.

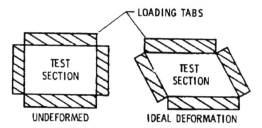

Figure 6.2. Undeformed and deformed configuration.

Comparing equations (5) and (2) it is possible to note that the Iosipescu test offers a greater reliability than the AFBT test. In fact, in both tests the mean shear stress evaluation is affected by errors connected to the measurement of geometric sizes. But in the AFBT test formula, two more parameters are present (a and b) than in the Iosipescu one.

All that was expounded for isotropic material can be extended for orthotropic ones. In this case if orthotropic ratio and notch depth assure a uniform stress distribution in the middle area, then still:

$$\tau_{xy} = \bar{\tau}_{xy} = \frac{P}{t(h-2d)}. \qquad (6)$$

Adams & Walrath (1987b) performed a finite element analysis to evaluate the stress fields that affect the notched part of specimens with orthotropic ratio E_x/E_y equal to 13.3 and 49.4.

They obtained shear stress concentration factors equal to 2.01 and 2.42 respectively. In the same way Herakovich & Bergner (1980) obtained a concentration factor equal to 1.95 for an orthotropic material ratio equal to 17.7. Adams & Walrath (1987b) deduced that the shear stress concentration effect could be reduced by increasing the angle and root radius of the notches. They observed that a 120° notch angle, a 0.2h depth and a 1.27 mm (0.05 in) root radius lead to a nearly uniform shear stress distribution in the notched section, but only for the lower orthotropic ratio material. In this case, equation (6) and numerically obtained strain values individuated a shear modulus with a 4.5% error. Instead, the shear stress concentration factor was still too high for the higher orthotropic ratio material. So it seems that good shear modulus estimations can be done using the mentioned specimen sizes, when the orthotropic ratio has moderate values. Nevertheless, test suitability depends on material nature and on the way it fails with regard to the stress components involved. Adams & Walrath (1987a) tested an AS4/3501-6 unidirectional composite (graphite/epoxy) using a specimen with the described sizes. They observed that material failure was preceded by two crack initiations at the notch vertexes.

These cracks propagated in the longitudinal direction until they became stable. After, at higher load values, other cracks initiated simultaneously in the area between notches until the specimen collapsed. The authors performed a finite element analysis of the specimen damaged by the first two cracks. They demonstrated that these two cracks did not influence the initial uniform shear stress distribution.

So the shear strength evaluated by equation (2) was significative (Adams & Walrath, 1984).

6   SHEAR TEST OF THIN PANEL

Ideally the shear test can be carried out on a rectangular or square thin panel whose edges are joined along all their length to the elements of a mechanism formed by four rigid rods. These are mutually hinged at their ends (Fig. 6.1). If the material is orthotropic, with two principal directions parallel to the panel edges, or if it is isotropic, by means of the mechanism it is possible to impose the displacements that correspond to a uniform shear stress distribution τ_{xy} in the panel. Really, it is impossible connecting panel edges and rods affecting only the transverse surfaces of the panel. Because of this, generally, specimens are cross-shaped with arms strengthened through very rigid tabs that are easily bolted to the rods (Fig. 6.2). In this way rod displacements are imposed to the boundaries of the central zone which would behave like a correctly connected panel.

Photoelastic analyses (Bryan, 1961, Liber et al. 1971) showed that specimens with this shape were affected by non-uniform stress fields in the measure zone. A shear stress concentration was observed along the edges as well as a normal stress concentration near the panel cor-

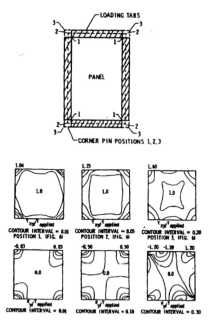

Figure 6.3. The effect of corner-pin location on panel stress distribution (Fairley & Baker, 1983).

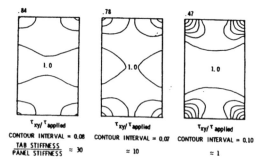

Figure 6.4. The effects of tabs stiffness on shear-stress distribution (Fairley & Baker, 1983).

ners. These effects produced specimen failure with wide lacerations along the edges next to the strengthening tabs. In the other case crack propagation and buckling in the diagonal direction occurred, depending on the sign of the normal stress acting in this part.

Fairley & Baker (1983) analysed the problem both numerically and experimentally using strain gages. They obtained a device, which generated a uniform shear strain distribution in a wide central area of the panel. They evaluated the acting stress fields by assuming that the strengthening tabs and rods were infinitely rigid. They considered three different ways in panel constraining with regard to the hinge position. Fig. 6.3 shows the stress field contour lines; curves are normalised with the mean value of the shear stress $\overline{\tau_{xy}}$ acting along the panel edges. When hinges are positioned on the panel vertexes, a nearly uniform shear stress field acts, meanwhile the tensile stress has zero value in the central part and has negligible values compared to the mean value $\overline{\tau_{xy}}$ in the corner areas. The authors also performed another finite element analysis to estimate the effects of the elasticity of the strengthening tabs. The results indicated that the in-plane shear stress distribution in the panel corners depends on the ratio of the in-plane stiffness of the panel to the stiffness of the loading tabs. Another important parameter is the distance from the corner pins to the nearest bolt on which the load is applied. Between the end of the tab and the nearest bolt, load transfer to the panel test section is through in-plane shearing along the loading tab. If the stiffness of the tab and that of the panel are comparable, in-plane bending

of the tab occurs between the corner and the nearest bolts. This causes a non-uniform stress distribution in the corner with an amplitude lower than in the central area (Fig. 6.4). If the tab stiffness is 30 times greater than the panel test section stiffness, the stresses are distributed more uniformly and there is a decrease in the extension of the area on which τ_{xy}^{max} acts. So the first situation is to be preferred in a shear test.

The panel should be sized to avoid buckling. If the geometry is fixed, the buckling load depends on the shear modulus, which is the quantity being measured. Thus in sizing the panel it is necessary to use an estimated value for G_{xy}. Unfortunately, there are no general graphical results for the buckling of a rectangular orthotropic plate with clamped edges. If the panel is square and simply supported, the critical shear stress value τ_{cr} can be obtained with the expression introduced by Davenport and Bert (1972):

$$\tau_{cr} = \frac{\pi^2 K D_y}{at^2}$$

where a is the edge length of the panel, t the thickness, D_y the transverse flexural stiffness of the panel and K is a dimensionless buckling coefficient which depends on D_y/D_x and $v_{xy}+(2D_{xy}/D_y)$.

7 FRACTURE TOUGHNESS TEST

Fracture toughness tests are not normalised yet for metal matrix composite materials. If the material is isotropic, the critical value of the mode I stress intensity factor can be determined according to the ASTM E 399 recommendations. This standard provides instructions about the identification of the failure plane and the failure direction with regards to the geometry of the specimen being tested. Furthermore the most suitable specimen configurations, its sizing, its processing and pre-cracking operations are described. The standard also fixes some dimensional limitations for the specimens in relation to the K_{Ic} measured value and the material tensile strength. For the C(T) specimen, the limitations are:

$$B \geq 2.5 \left(\frac{K_{lc}}{\sigma_{ys}} \right)^2 ,$$

$$a \geq 2.5 \left(\frac{K_{lc}}{\sigma_{ys}} \right)^2 ,$$

where B is the specimen thickness and a is the crack length. If these limitations are respected, a plane strain state is present in the specimen and the instability condition occurs without significant plastic flows.

Wu (1963) and Sih et al. (1965) analysed stress fields acting in proximity to the crack tip of orthotropic material specimens. They used the Lekhnitskii's (1963) extension to the anisotropic case of the Muskevilshli's (1963) method for the isotropic problem. Later Wu (1967) demonstrated that with any asymptotic stress condition, the stress distribution near the crack tip can be always decomposed into the symmetric and antisymmetric part. Liebowitz (1968), Hahn (1970) and Corten (1972) obtained analytical solutions for particular specimen shapes and load conditions. For these cases mathematical expressions of the stress intensity factors are available so that it is possible to evaluate their critical values experimentally. The amplitude of the plastic zone in the condition of incipient unstable propagation can be approximately evaluated by means of the most suitable yielding criterion for the given material. In the literature no indication was found about the specimen thickness which realises the maximum yielding effect at the crack tip.

8 FATIGUE TEST

No normalisation is available for fatigue tests on metal matrix composites, as for the toughness ones. ASTM E 466 contains instructions about the specimen preparation, which can be also used in uniaxial fatigue tests on isotropic composites. Instead, if uniaxial tests in the principal directions of anisotropic material are to be performed, specimens can be prepared as those used in tensile tests. In order to reduce the stress concentration effect in the transition zones between the useful length and the gripping zones, the fillet radius must be maximised.

The specimen has to be machined without localised overheating or hardening of its surface or in the underlying material layers. Otherwise cracks could already initiate before the load application.

It is a good thing to remember that the British Standards Institution provided several recommendations for specimen machining. ASTM (1974) published these instructions. Furthermore, in the appendix of the above-mentioned ASTM standard, there is reported a series of typical machining details which involve specimens realised in traditional metals. This standard advises to check the specimen surface with the naked eye or with a lens that has magnification capability lower than 20x. If irregularities are discovered like cracks, roughing marks and scorings, then the specimen is unsable. But recently De Iorio et al. (1998a, 1998b) have observed that specimen geometry, its superficial finishes and experimental procedures must be chosen depending on the material being tested in order to characterise it correctly. Furthermore they noted that standards defined accurately the working sequence that leads to the final specimen geometry for isotropic and homogeneous materials; but this attention could be useless if the material, like a sintered composite, contains microcavities. In fact if these are present in proximity to the surface, they affect the material strength to fatigue loads more than the working marks do.

9 CONCLUSIONS

In this work an analysis of the experimental procedures for the composite mechanical characterisation was illustrated, particularly for the metal matrix ones. Modalities and fixtures of every test were described as well as the analytical relations used in the experimental data analysis.

For uniaxial tests in the principal directions, some expressions were provided which can be used when the specimen geometry established by standards is not suitable. Furthermore, several devices were discussed in order to avoid undesired phenomena which can invalidate the test results.

The off-axis tests produced the most important results and relating experimental strategies the research has obtained to date.

For the shear properties evaluation, three kinds of test were illustrated: Iosipescu test, AFBT test and shear panel picture test. The first two use specimens with the same geometry. Examining the analytical relations used in the experimental data analysis, it was observed that the Iosipescu test is preferred to the AFBT test, even if the latter uses simpler fixtures. The third kind of test gives more accurate evaluations of the shear modulus and of the shear stress critical value; but this is applied in cases in which the first two can not be used, because of its higher cost.

Finally, the evaluation methods of the mode I and II fracture toughness were described together with the more suitable procedures used in the estimation of the composites fatigue strength.

REFERENCES

Adams F., Walrath D.E. 1987a, "Current Status of the Iosipescu Method", Comp. Mat., vol. 21, pp. 494-507.

Adams F., Walrath D.E. 1987b, "Further Development of the Iosipescu Shear Test Method", Exp. Mech., vol. 31, pp. 113-119.

Adams F., Walrath D.E. 1984, "Verification and Application of the Iosipescu Shear Test Method", Report No. UWME-DR-401-103-1, Department of Mechanical Engineering, University of Wyoming (NASA GRANT No. NAG-1-272).

Bryan E.L. 1961, ASTM STP 289, pp. 90-94.

ASTM Standard and Literature Reference for Composite Materials 1987, American Society for Testing and Materials, PA.

ASTM D3552-82, "Standard Test Method for Tensile Properties of Fiber-Reinforced Metal Matrix Materials".

ASTM E 399, "Standard Test Method for Plane Strain Fracture Toughness of Metallic Materials".

ASTM E 466, "Standard Practice for Conducting Force Controlled Constant Amplitude Axial Fatigue Test of Metallic Materials".

ASTM E 8M, "Standard Test Method for Tension Testing of Metallic Materials".

ASTM E83, "Standard Practice for Verification and Classification of Extensometers".

Corten H.T. 1972, in "Fracture" (H. Liebowitz, ed.), vol. 7, Academic Press, New York.

Davenport O.B., Bert C.W. 1972, J. Aircr., vol. 9, p. 477.

De Iorio A., Ianniello D., Iannuzzi R., Penta F. 1998a, "Practice for Conducting Fatigue Test at Various Temperatures for a Metal Matrix. Composite Material", European Conference on Composite Materials, Naples, Italy.

De Iorio A., Ianniello D., Iannuzzi R., Penta F., Florio G., Scalabrino 1998b, G., "Preparation Procedure of Fatigue Test MMC_p Specimens", 4° Seminar on Experimental Techniques and Design In Composites, University of Sheffield, England.

Fairley G.L., Baker D.J. 1983, "In-Plane Shear Test of Thin Plates", Exp. Mech., vol. 22, pp. 81-88.

Hahn H.G. 1970, VDI-Forschungshelft 542, VDI-Verlag, Berlin.

"Handbook of Fatigue Testing" Oct. 1974, ASTM - STP566.

Herakovich C.T., Bergner H.W. 1980, "Finite Element Analysis of Notched Coupon Specimen for In-Plane Shear Behaviour of Composites", pp. 149-154.

Iosipescu N., 1967, "New Accurate Procedure for Single Shear Testing of Metals", J. Mat'ls, vol. 2, pp. 537-566.

Iosipescu N., 1963, "Photoelastic Investigations of an Accurate Procedure for the Pure Shear Testing of Materials", Rev. De Mec. Appl., vol. 1, p. 145.

Liber T., Daniel I.M., Ahimaz F.J. 1971, 2° Conf. Compos. Mater.: Testing & Design, Anaheim, California.

Kreider, K.G. 1969, "Mechanical Testing of Metal Matrix Composites", Composite Materials: Testing and Design, ASTM STP460, pp. 203-214.

Lekhniskii S.G. 1963, "Theory of Elasticity of an Anisotropic Elastic Body (English Translation by P. Fern), Holden-Day, San Francisco.

Liebowitz G.C., 1968, in "Fracture" (H. Liebowitz, ed.), vol. 2, Academic Press, New York and London.

Muskhelishvili N.I. 1963, "Some Basic Problems of the Mathematical Theory of Elasticity", P. Noordhoff L.t.d., Groningen.

Pagano N.J., Halpin J.C. 1968, "Influence of End Constraint in the Testing of Anisotropic Bodies", J. Comp. Mat., vol. 2, p. 18.

Richards G.L., Airhart T.P., Ashton J.E. 1969, "Off-Axis Tensile Coupon Testing", J. Comp. Mat., vol. 3, p. 586.

Rizzo R.R. 1969, "More on the Influence od End Constraints on Off-Axis Tensile Test", J. Comp. Mat., No. 3, p. 302.

Roebuck B., Hayes D.M., Cooper P.M., Murray, S.N. 1987, "Tensile Test on SiC Particulate Reinforced Aluminium Matrix Composite", Report DMA(D) 585, NPL, Teddington.

Sih G.C., Paris P.C., Irwin G.R. 1965, Int. J. Fracture Mech 1, 3, p. 189.

Slepetz, J.M., Zagaeski T.F., Novello R.F. 1978, "In Plane Test for Composite Materials", Report No. AMMRC TR 78-30, Army Materials and Mechanics Research Center, Watertown, MA.

Sullivan J.L., Kao B.G., Van Oene, H. 1984, "Shear Properties and a Stress Analysis Obtained from Vinyl-Ester Iosipescu Specimens", Exp. Mech., vol. 24, pp. 223-232.

Wu E.M. 1967, in "Composite Materials Workshop" (S.W. Tsai, J.C. Halpin, N.J. Pagano, eds), Technomic Publ., Stamford, Connecticut.

Wu E.M. 1963, "On the application of Fracture Mechanics to Orthotropic Plates", TAM Rep. No. 248, Univ. Of Illinois, Urbena.

Wu E.M., Thomas R.L. 1986, "Off-Axis Test of a Composite", J. Comp. Mat., vol. 2, No. 4, p. 523.

Experimental Techniques and Design In Composite Materials 4, Found (Ed.)
© 2002 Swets & Zeitlinger, Lisse, ISBN 90 5809 370 0

Correlation of the Mechanical Properties of Advanced Composites with their Interfacial Performance through Single Fibre Fragmentation Testing

A. Abu Bakar & F.R. Jones
Department of Engineering Materials, Sir Robert Hadfield Building
The University of Sheffield, Mappin Street S1 3JD, U.K.

ABSTRACT: An advanced epoxy resin used in commercial applications has been prepregged with carbon fibre with differing surface finish. The interface dominant mechanical properties of the unidirectional materials have been assessed. Thus, interlaminar shear strength (ILSS), transverse strength and modulus as well as unidirectional tensile strength and modulus have all been measured. Single fibre fragmentation testing has been carried out using the same fibre and resin combinations. The relationship between interface quality and composite performance is attempted. It was found that the correlation between the properties of microcomposite and macrocomposite is good.

1 INTRODUCTION

There has been increasing interest in being able to understand the physical and chemical mechanisms responsible for fibre-matrix adhesion as well as a role of fibre-matrix adhesion on advanced composites properties. Many earlier works on the development of composite materials considered fibre-matrix adhesion to be a necessary condition to ensure good composites properties. The majority of effort was concentrated on increasing fibre-matrix adhesion through the use of surface treatments and sizings on fibres (Cheng et al. 1994, Yumitori et al. 1994, Cheng et al. 1993, Cheng 1994).

It is desirable to have a testing method to measure the adhesion between fibre and matrix that can provide a reproducible and reliable method for investigating and measuring fibre-matrix adhesion. Several testing methods have been developed for measuring fibre-matrix adhesion using single fibre or a bundle of fibres. The aim was to measure fibre-matrix adhesion in away that would be a predictor of fibre-matrix adhesion as well as to understand the fundamental physics and chemistry of adhesion. There are three basic measurement methods that have been used extensively to measure fibre-matrix adhesion, namely, single fibre fragmentation test, fibre pull-out test and fibre micro-indentation test.

The single fibre fragmentation test is now commonly used for measuring fibre-matrix adhesion of fibre reinforced composites due to ease of producing the sample and provide much information. In this method, the fibre is embedded in a matrix coupon, a tensile load is applied to the coupon and interfacial shear stress mechanism is relied upon to transfer the coupon tensile forces to the embedded fibre through the interface (Kelly & Tyson

1965) As the load is increased on the specimen, shear forces are transmitted to the fibre along the interface. The fibre tensile stress increases to the point where the fibre fracture strength is exceeded and the fibre breaks inside the matrix. The fibre continues to break into shorter lengths until the fragment lengths become too short to break any further.

The interfacial shear strength is calculated from the fragmentation test by assuming that the interfacial shear stress is constant over the fragment length either due to matrix flow in a complete plastic matrix or due to frictional stress transfer as a result of interfacial debonding. However, the results obtained with conventional data reduction techniques based on the constant shear model of Kelly-Tyson (Kelly & Tyson 1965) or the partial debonding of Piggott (Piggott 1978) have several limitations. Different micromechanical features such as matrix yielding, transverse matrix cracking and interfacial debonding play an important role in the overall characterisation of the interface and are not properly accounted for in the existing data reduction techniques. This complexity in the fragmentation test makes correlation with the fibre surface chemistry and mechanical properties of the macrocomposites difficult (Drzal & Madhukar 1993, Ivens et al. 1993, Hoecker & Karger-Kocsis 1994, Cheng et al. 1994). During the micromechanical testing of fibre reinforced composites, it is assumed that a particular trend in the value of interfacial shear strength of the fibre with different surface treatments will be also reflected by the macromechanical testing. This correlation however, is difficult to prove and sometimes does not hold true. It was shown that different macromechanical properties show different sensitivity towards the fibre-matrix adhesion measured using the fragmentation test.

The object of this paper is to try to correlate the mechanical properties of real composites with their interfacial performance through single fibre fragmentation testing using a Cumulative Stress Transfer Function (CSTF) technique proposed by Tripathi and Jones (Tripathi & Jones 1996). In this technique, the interfacial shear stress associated with the tensile stress in the individual fibre fragments are predicted from the plasticity effect model and the total tensile stress transferred to all of the fibre fragments at a particular matrix strain is calculated. The details of CSTF technique is given elsewhere (Tripathi 1995, Tripathi & Jones 1997) Since, CSTF technique is a direct measure of the efficiency of the stress transfer to the fibre across the interface, the correlation of the single fibre fragmentation (micromechanical) test with the macromechanical tests is thought to be easier.

2 EXPERIMENTAL PROCEDURE

2.1 *The material*

2.1.1 *Carbon fibre*
HTA5131 and HTA5001 carbon fibres were supplied by Toho Carbon Fibre Inc. HTA5131 is a surface treated and sized (TS) and HTA5001 is a treated and unsized (TU) carbon fibre manufactured from polyacrylonitrile precursor.

2.1.2 *Epoxy resin*
MTM60 epoxy resin from Advanced Composites Group Ltd. (ACG) was used for the single fibre fragmentation test. It is a medium temperature curing resin whose prepregs are mainly used for manufacturing structural components in motor sports. The resin was degassed in the vacuum oven before it was used in preparing the fragmentation test samples. The following cure cycle was used: 60°C for 30 minutes; 80°C to 120°C at 20°C per hour; 90 minutes at 120°C, followed by natural cooling to room temperature. The cured resin had a distinctive light yellow colour and is semi-transparent.

2.2 *Single fibre fragmentation test*

Single fibres were extracted at random from the tow, mounted on a steel wire, mounted on PTFT mould and embedded in a hot cured MTM60 resin. Care was taken to ensure that the fibre under test has not been touched and the fibre remains straight and aligned. The cured specimens were then ground and carefully polished. The final dimensions of the specimens were 80 × 9.8 × 1.65 mm. Test specimens were subjected to uniaxial tension at a displacement rate of 0.13 mm/min. The fragmentation of single fibre was continuously monitored across the specimen length. The test was stopped at every one percent increment of applied strain, starting from 4% and all the fragments in the specimen were digitised to the pictures. The digitisation of the whole specimen length normally took less than 5 minutes. The fibre fragments, which could not be accommodated in

the video screen for the digitisation of the picture (typically greater than 1.50 mm) were directly measured using a digital vernier calliper fitted to the microscope. The test was stopped after the tensile failure of the specimen.

2.3 *Prepregs manufacturing*

A laboratory drum winding machine was used to produce prepregs with different type of carbon fibres. MTM60 resin film with 50 gram per square metre (gsm) weight was attached to the drum roller and carbon fibres were wound around the drum roller. The drum roller and bath speed were set to obtain ~ 60% fibre volume fraction prepregs. The drum roller temperature was set at 40°C. When finished, the prepreg was cut from the drum roller and another layer of resin film was placed on top of the prepreg. The prepreg was then rolled on a bench by hand using a roller.

2.4 *Laminates manufacturing*

The prepregs were vacuum bagged prior to curing in an autoclave. The vacuum bag technique was used to improve the wetting of the fibres by the matrix. The cure cycle for curing the laminate in the autoclave is the same as the cure cycle for preparing single fibre fragmentation test specimen.

2.5 *Mechanical test*

All the specimens for mechanical testing were cut from the laminate plate using diamond-coated cutting wheel. Longitudinal tensile test was done according to BS 2782: Part 3: Method 320E: 1976, transverse tensile test according to ASTM D3039 (1976) and short beam shear test according to ASTM D2344 (1984). All the tests were performed using Mayes Universal Testing Machine model SM200.

3 RESULTS AND DISCUSSION

3.1 *Single fibre fragmentation test*

The mechanical properties of MTM60 resin for the fragmentation test are shown in Table 1.

Table 1. Mechanical properties of the MTM60 epoxy resin for the fragmentation test.

Mechanical property	Value
Tensile yield strength (MPa)	81.52 ± 2.34
Shear yield strength (MPa)	47.07 ± 2.34
Tensile modulus (GPa)	3.20 ± 0.24
Failure strain (%)	7.83 ± 0.26

The machine was stopped for about 5 minutes at every 1% applied strain to digitise the pictures of the fragment lengths. During this period, the tensile stress

Table 2. Fibre strength data for different carbon fibres used in the fragmentation test.

Property	TS	TU
Average fibre diameter (μm)	7.25 ± 0.25	7.20 ± 0.21
Average fibre strength at 6.25 mm length (GPa)	3.56 ± 0.78	3.84 ± 0.75
Elastic modulus of the fibre (GPa)	238	238
Weibull modulus	5.10	5.74

Table 3. Material properties used for the calculation of CSFT value.

Material property	Value
Transverse modulus of fibre (GPa)	14
Longitudinal Poisson's ratio of fibre	0.20
Transverse Poisson's ratio of fibre	0.25
Longitudinal shear modulus of fibre (GPa)	20
Friction coefficient	0.2
Fibre volume fraction	0.001
Longitudinal thermal expansion coefficient of fibre	-0.1×10^{-6}
Transverse thermal coefficient expansion of fibre	18×10^{-6}
Thermal expansion coefficient of matrix	40×10^{-6}

Table 4. Fragmentation test results for treated-sized carbon fibre at different applied strains (standard deviation in brackets).

Strain (%)	4	5	6	7	8
Average fragment length (mm)	0.68 (1.13)	0.34 (0.17)	0.27 (0.08)	0.25 (0.07)	0.23 (0.06)
Critical fibre Length (mm)	0.91	0.45	0.36	0.33	0.31
Number of fragment per 10 mm	14.7	28.7	37.3	40.6	43.2
Apparent IFSS (MPa)	20.78	47.62	62.74	68.78	76.00
CSTF value (MPa m/m)	2974	2254	1771	1629	1533

in the sample relaxed by about 5 MPa. However, this relaxation quickly disappeared as test continued and the tensile stress curve resumed it original trend. The other fibre and matrix properties used for the calculations of CSTF value are given in Table 2 and 3.

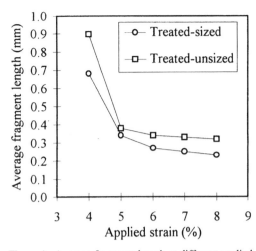

Figure 1. Average fragment length at different applied strains obtained during the fragmentation of single fibre composites with different interfaces.

3.1.1 Treated-sized fibre

The single fibre fragmentation test results for treated-sized fibre are shown in Table 4. During fragmentation test of MTM60/treated sized carbon fibres, it was observed that the first fibre fracture occurred at an applied strain greater than 2.6%. There were very few fibre fragments shorter than 1.50 mm at applied strain less than 3%. The first detail measurements of fibre fragment lengths were carried out at an applied strain of 4%. Only a few fibres longer than 1.50 mm exist at 4% strain that breaks quickly into smaller fragments as the applied strain is further increased. A broader distribution in the fragment length from sample to sample was observed at 4% applied strain which narrowed as the applied strain increased.

As the applied strain increased, the number of fibre fragments increased and the average fragment length decreased (Figure 1). It may seem that the fragmentation process is not reaching saturation. However, the sum of all fragment lengths is also increasing with the applied strain. In other words, an increased region of the single fibre fragmentation test specimen is being used for the measurement of fragment lengths. Consequently, the true measure of the fragment saturation will be the number of fragments per unit length of the test specimen. We have assumed that a unit length of 10 mm for this purpose. It can be seen from Table 3 and Figure 2 that the number of fragments per 10 mm increases with the increase of applied strain. From these results it can be concluded that saturation process is not reached for treated-sized fibre. No interfacial debonding was observed during the fragmentation test of treated-sized fibre specimens.

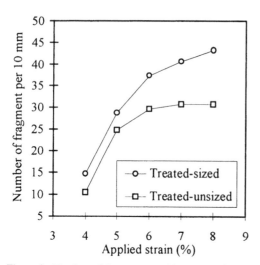

Figure 2. Number of fragment per 10 mm at different applied strain obtained during the fragmentation of single fibre composites with different interfaces.

Table 5. Fragmentation test results for treated-unsized carbon fibre at different applied strains (standard deviation in brackets)

Strain (%)	4	5	6	7	8
Average fragment length (mm)	0.90 (1.03)	0.38 (0.15)	0.34 (0.11)	0.33 (0.09)	0.32 (0.10)
Critical fibre length (mm)	1.20	0.51	0.45	0.44	0.43
Average debonding length (mm)	0.004 (0.01)	0.03 (0.03)	0.04 (0.02)	0.06 (0.02)	0.08 (0.04)
Number of fragment per 10 mm	10.5	24.8	29.7	30.7	30.8
Apparent IFSS (MPa)	10.16	33.77	48.16	49.88	51.72
CSTF value (MPa m/m)	3863	2410	1796	1544	1375

3.1.2 Treated-unsized fibre

A fragmentation profile, similar to treated-sized fibre was observed for the treated-unsized fibre as shown in Table 5. However, in the case of treated-unsized fibre, the onset of the fragmentation began at higher applied strain that is at around 3.0%. Consequently fewer fragments were observed at an applied strain of 4% in comparison to treated-sized fibre.

Further, similar to the fragmentation of treated-sized fibre, a broader distribution of the fragment length from sample to sample was observed at 4% applied strain

which become narrower as the applied strain increased. However, at 4% and 5% applied strain, the fragment length distribution was narrower but at 6%, 7% and 8% the fragment length distribution was broader for treated-unsized fibre than that in the case of treated-sized fibre. Interfacial debonding was observed during the fragmentation test of treated-unsized fibre specimens and increased with applied strain. It can be seen in Table 5 and Figure 2 that the number of fragments per 10 mm reached a plateau at 6% strain. So, it can be concluded that the saturation process is reached at about 6% strain in treated-unsized fibre specimen.

It can be seen in Table 4 and 5 that the average fragment length of treated-sized fibres is shorter than that of treated-unsized fibres. We can also see that average fragment aspect ratio of treated-unsized fibres at all applied strain is higher than that of treated-sized. Interfacial debonding also is observed in treated-unsized fibre specimens. Hence, it can be concluded that the fibre-matrix interface in the case of treated-sized fibres is better than that in treated-unsized fibres.

3.1.3 Interfacial shear strength and CSTF value

The single fibre fragmentation test results for treated-sized and treated-unsized fibres were used to calculate the fibre-matrix interfact parameter, using Kelly-Tyson equation and Cumulative Stress Transfer Function (CSTF) methodology at different applied strain and the results are shown in Table 4 and 5, respectively. The details of the properties used for the CSTF calculation are given in Table 1, 2 and 3. Fibre fragments longer than 1.50 mm (aspect ratio > 200) were excluded from the calculation of CSTF value since these fragments will essentially behave as a continuous fibre.

It can be seen from Table 4 and Table 5 that the critical fibre length decreases as the applied strain increases since the average fragment length decreases as well and the relationship between the critical fibre length and average fragment length is linear. Fibre strength at the critical fibre length increases as the critical fibre length decreases since fibre strength is inversely proportional to the fibre length.

The interfacial shear strength and CSTF values for the treated-sized fibre is given in Table 4. It can be seen in Figure 3 that the value of interfacial shear strength increases as the applied strain increases. This is expected since one of the major assumption for the calculation of interfacial shear strength value from Kelly-Tyson equation is that a better interface leads to shorter fragment length. However, the values of interfacial shear strength at applies strain less than 5% for treated-unsized fibre is meaningless since the saturation in the fibre fragmentation process is not achieved. On the other hand, CSTF value decreases as the applied strain increases as shown in Figure 4. This is in accordance to the assumption for the calculation of CSTF value that assumes that a shorter fibre is subjected to lower levels of stress in comparison to longer fibre. As the fibre fragments become shorter with the increase in strain, CSTF value will decrease. The fragmentation test results for

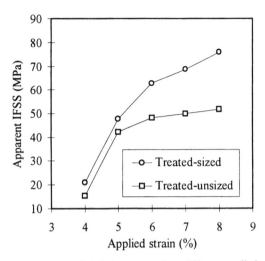

Figure 3. Interfacial shear strength at different applied strains obtained during the fragmentation of single fibre composites with different interfaces.

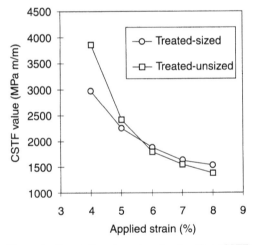

Figure 4. Cumulative stress transfer function (CSTF) values at different applied strains obtained during the fragmentation of single fibre composites with different interfaces.

the treated-unsized fibre show a similar trend.

A comparison of the interfacial shear strength for the treated-sized and treated-unsized fibres at similar applied strain shows that treated-sized fibre has higher interfacial shear strength compared to treated-unsized fibre. The prediction of inferfacial quality by Kelly-Tyson model is acceptable since treated-sized fibre shows shorter average fragment length in comparison with treated-unsized fibre. However, since treated-sized fibre is found not to achieved saturation, the interfacial

shear strength value for treated-sized fibre is meaningless. It can be seen that the interfacial shear strength obtained from constant-shear model at more than 5% strain is higher than the shear yield strength of the matrix for treated-unsized fibre. The shear yield strength of the matrix calculated based on von Mises criterion is 47.07 MPa. Since interfacial debonding was observed during the fragmentation test, it is expected that the interfacial shear strength should be lower than the shear yield strength of the matrix. This is one of the major limitations of the Kelly-Tyson interfacial shear strength and has been highlighted by several other workers (Melanitis et al. 1992, Feillard et al. 1994, Tripathi et al. 1996 & Tripathi & Jones 1997).

Meanwhile, the comparison of CSTF value shows that treated-sized fibre has higher CSTF value at 6% to 8% applied strain compared to treated-unsized fibre. These results are in accordance to the assumption of calculation of CSTF value. CSTF technique assumes that longer fibres are subject to higher level of stress in comparison to shorter fibres. As treated-unsized fibre has a longer average fragment length especially at 4 and 5% strain, it will has higher CSTF value. However, the presence of interfacial debonding will reduced the CSTF value. As we can see that at 6 to 8% strain, interfacial debonding in treated-unsized fibre specimen has significant effect of CSTF value.

These results show good agreement between CSTF value with fracture information obtained from the fragmentation test. Hence, it can be concluded that the prediction of interfacial adhesion using the CSTF technique gives more reliable and accurate results. Further, CSTF value shows the same trend in the quality of the interface irrespective of the applied strain and/or the saturation during the fragmentation test. Hence, the prediction of the quality of the interface at an arbitrary applied strain can be made provided sufficient number of fragments are obtained during the fragmentation test at particular strain.

3.2 *Mechanical test*

The mechanical test results for all the composites are shown in Table 7. We can see that the mechanical properties of MTM60/TS composite are slightly higher than that of MTM60/TU composite except for transverse tensile strength that is significantly higher.

This shows that some mechanical properties are insensitive to differences in the quality of fibre-matrix adhesion in terms of value of mechanical properties.

A comparison of longitudinal modulus of the two composites shows that the difference is not significant (Figure 5). The insensitivity of the longitudinal tensile modulus to fibre-matrix adhesion quality should be expected, considering that in longitudinal tension specimen, fibre and matrix are connected through the interface in parallel. Most of the applied load is carried by the longitudinal fibres. The role of interface and matrix is limited to transferring stress from highly stressed fibres to the neighbouring fibres carrying relatively low

Table 7. Mechanical properties of MTM60/carbon fibre composites (standard deviation in brackets)

Property	TS	TU
Tensile strength (MPa)	1614.68 (80.31)	1496.53 (47.81)
Tensile modulus (GPa)	163.02 (9.16)	158.30 (13.07)
Transverse strength (MPa)	52.51 (3.89)	39.29 (5.74)
Transverse modulus (GPa)	9.53 (0.82)	9.28 (1.05)
ILSS (MPa)	66.33 (2.47)	63.11 (3.19)

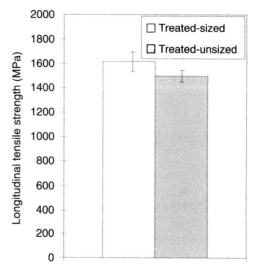

Figure 6. Longitudinal tensile strength of treated-sized and treated-unsized composites.

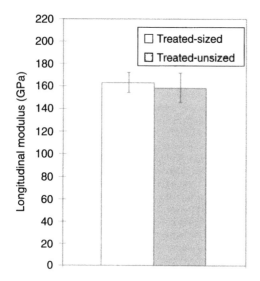

Figure 5. Longitudinal modulus of treated-sized and treated-unsized composites.

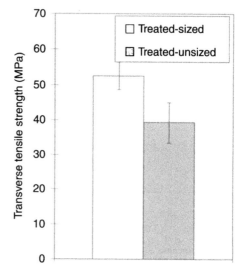

Figure 7. Transverse tensile strength of treated-sized and treated-unsized composites.

stress so as to result in a uniform stress distribution in the composite.

Although the longitudinal tensile strengths of both laminates as shown in Figure 6 are not much different, the failure modes are found to be different. MTM60/TS composite show cracks have cut across the specimen with little fibre-matrix splitting. This type of failure is characterised as primarily the matrix failure or brittle failure. The same observation is also reported earlier (Drzal & Madhukar 1993, Ivens et al. 1994). Meanwhile, the failure of MTM60/TU composite is like an explosion resulting in a 'brush-like' appearance of the fractured specimen that show extensive fibre-matrix splitting. This type of matrix cracking for fibre with lower interface strength was also previously reported by several workers (Drzal & Madhukar 1993, Bader 1988, Ivens et al. 1994).

For transverse tensile strength, where failure mechanism is dominated by matrix and interface, the influence of the interface is very clear. MTM60/TS composite has higher transverse tensile strength compared to MTM60/TU composite. Like longitudinal tensile modulus, the transverse tensile modulus is also insensitive to the fibre-matrix adhesion. The results for transverse tensile strength and modulus shown in Table 7, Figure 7

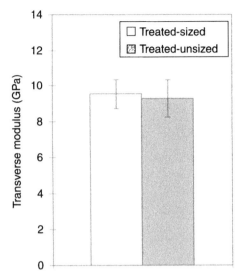

Figure 8. Transverse modulus of treated-sized and treated-unsized composites.

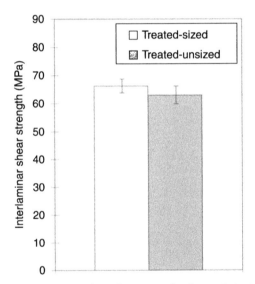

Figure 9. Interlaminar shear strength of treated-sized and treated-unsized composites.

and Figure 8 correlate well with those reported earlier by several workers (Drzal & Madhukar 1993, Hoecker & Karger-Kocsis 1994).

Interlaminar shear strength (ILSS) results from short beam shear test are shown in Table 7 and Figure 9. It seems that the interlaminar shear strength is insensitive to the change in fibre-matrix adhesion. This is unexpected since short beam shear test has been established as standard for assessing the quality of fibre-matrix

adhesion especially for thermoset composites. It was also found (Hocker & Karger-Kocsis 1994) that interlaminar shear strength is insensitive to fibre-matrix adhesion. Another worker (Drzal & Madhukar 1993) also found only a slight improvement in interlaminar shear strength from treated-unsized to treated-sized composites.

Both MTM60/TS and MTM60/TU composite specimens show a single crack at the side of the specimen that propagated from near the centre of the specimen right across one half (Wisnom 1994). Both specimens also show a crack that is considered as an irregular crack (Daniels et al. 1971) and also delamination in several planes across the specimen depth.

We can see from macrocomposite properties results that a treated-sized composite has better properties compared to a treated-unsized composite. These results correlate well with the fragmentation test results which show that treated-sized fibre has a higher CSTF value than that of treated-unsized fibre. Interfacial debonding is also observed in treated-unsized fibre which shows it has a lower fibre-matrix adhesion.

4 CONCLUSION

From the results obtained in this paper, a few conclusions can be drawn.

1. Treated-sized fibre shows better fibre/matrix adhesion compared to treated-unsized fibre as shown in CSTF value and failure mode during fragmentation test. This finding is confirmed by the mechanical properties of both composites.
2. CSTF technique was found to be a reliable and accurate method to differentiate the differences in interface quality in the fibres regardless of saturation process in fragmentation test specimen.
3. The increase in mechanical properties of macrocomposites is significant with the increase of fibre/matrix adhesion in microcomposites.

5 ACKNOWLEDGEMENT

AAB would like to thanks the Malaysian Government for the financial support and Advanced Composites Group Ltd for the supply of resin and the use of laboratory drum winding machine.

REFERENCES

Bader M.G. 1988. Tensile strength of uniaxial composites. *Sci. Eng. Comp. Mater.* 1: 1-11.
Cheng T.H., Jones F.R. & Wang D. 1993. Effect of fibre conditioning on the interfacial shear strength of glass-fibre composites. *Comp. Sci. Tech.* 48: 89-96.
Cheng T.H. 1994. The role of sizing resin on the micromechanics of fibre composites. *Ph.D. thesis.* The University of Sheffield, U.K.
Cheng T.H., Zhang J., Yumitori S, Jones F.R. & Anderson C.W. 1994. Sizing resin structure and interphase formation in carbon fibre composites.

Composites. 25(7): 661-670.

Daniels B.K., Harakas N.K. & Jackson R.C. 1971. Short beam shear tests of graphite fibre composites. *Fibre Science and Technology*. 3: 187-208.

Drzal L.T. & Madhukar M. 1993. Fibre-matrix adhesion and its relationship to composite mechanical properties. *J. Mater. Sci*. 28: 569-610.

Feillard P., Desarmot G. & Favre J.P. 1994. Theoretical aspects of the fragmentation test. *Comp. Sci. Tech*. 50: 265-279.

Hoecker F. & Karger-Kocsis J. 1994. Effect of the interfaces on the mechanical response of CF/EP microcomposites and macrocomposites. *Composites*. 25(7): 729-738.

Ivens J., Wevers M & Verpoest I. 1994. Influence of carbon fibre surface treatment on composite UD strength. *Composites*. 25(7): 722-728.

Kelly A. & Tyson W.R. 1965. Tensile properties of fibre reinforced metals: copper/tungsten and copper/molybdenum. *J. Mech. Phy. Solids*. 13: 329-350.

Melanitis N., Galiotis C., Tetlow P.L. & Davies C.K.L. 1992. Interfacial stress distribution in model composites Part 2: Fragmentation studies on carbon fibre epoxy system. *J. Comp. Mater*. 26: 574-610.

Piggott M.R. 1978. Expressions governing stress-strain curves in short fibre reinforced polymers. *J. Mater. Sci*. 13: 1709-1716.

Tripathi D. 1995. Micromechanics of interfaces in short fibre reinforced polymer composites. *Ph.D Thesis*. The University of Sheffield, U.K.

Tripathi D., Chen F. & Jones F.R. 1996. The effect of matrix plasticity on the stress fields in a single filament composite and the value of interfacial shear strength obtained from the fragmentation test. *Proceeding of the Royal Society A: Mathematical and Physical Sciences*. 452: 621-653.

Tripathi D. & Jones F.R. 1997. Measurement of load bearing capability of the fibre/matrix interface by single fibre fragmentation. *Comp. Sci. Tech*. 57: 925-935.

Wisnom M.R. 1994. Modelling of stable and unstable fracture of short beam shear specimens. *Composites*. 25(6): 394-400.

Yumitori S, Wang D. & Jones F.R. 1994. The role of sizing resins in carbon fibre-reinforced polyethersulfone (PES). *Composites*. 25(7): 698-705.

Section 3: *Design*

Experimental Techniques and Design In Composite Materials 4, Found (Ed.)
© 2002 Swets & Zeitlinger, Lisse, ISBN 90 5809 370 0

Numerical static and dynamic analysis of a SMC-R bumper for industrial trucks

D. Bragiè
Dipartimento di Meccanica, Politecnico di Torino, Italia

F. Curà
Dipartimento di Meccanica, Politecnico di Torino, Italia

G. Curti
Dipartimento di Meccanica, Politecnico di Torino, Italia

E. Indino
Centro Ricerche Fiat, Orbassano, Italia

ABSTRACT: The subject of this paper is a numerical study of an SMC-R bumper for industrial trucks. The numerical simulation of the moulding process (Plastec Code) was first performed to obtain the orthotropic mechanical properties and the glass fibres orientation. By using a Finite Element Code, static and dynamic analyses were then carried out on a 8000 elements FEM model with orthotropic characteristics. The static study was performed by employing established working loads. The dynamic analysis was carried out in both free-free and clamped conditions. The numerical results were compared both to those available in literature obtained by using an isotropic FEM model and to those obtained experimentally on the real bumper.

1 INTRODUCTION

SMC (Sheet Moulding Compound) is a generic name for composite materials consisting of glass fibres filled in a thermoset polymer matrix produced in sheet form.

In the so called SMC-R (Sheet Moulding Compound Random) the reinforcement is by means of random disposed glass fibres (the maximum fibre length is 50 mm).

The mechanical properties of the SMC are greatly influenced by the moulding process of the component. During moulding the flow of the melting material causes the fibres to assume preferential orientation in the different zones of the mould and the properties of the material can become highly anisotropic.

The aim of this work is to carry out common research developed with other Italian universities (Universita di Cagliari, Universita di Catania and Universita di Padova), based on preliminary studies performed at IVECO and Centro Ricerche Fiat, in order to evaluate the reliability of experimental and numerical methods in the analysis of complex shaped composite structures.

The object of this paper is the SMC bumper IVECO SPR1 60-100 N.P. (see Figure 1) for EUROCARGO industrial trucks (medium size vehicles IVECO) made of SMC-R33 (33% by weight of glass fibres).

Cura et al. (1996) performed the experimental static analysis on the above quoted bumper and the corresponding numerical one by means of an isotropic FEM model.

Curti et al. (1996) performed the experimental dynamic analysis in both free-free and clamped conditions.

Bragie et al. (1997) utilized an isotropic FEM model in order to study both the static and the dynamic behaviour of the bumper; this isotropic model, developed by Cura et al. (1996), was then modified according to the actual values of the thickness of the bumper.

Lazzarin et al. (1996) performed the numerical simulation of the moulding process of the bumper to obtain the fibre orientation distribution and the corresponding mechanical properties.

The aim of this work is to perform both a static and a dynamic analysis on the bumper by means of an anisotropic FEM model; the mechanical properties of the component are obtained from a numerical simulation of the moulding process (Plastec Code).

Furthermore the numerical results are compared both to those available in literature obtained by using an isotropic FEM model and to those obtained experimentally on the real bumper.

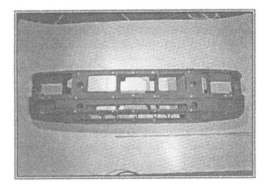

Figure 1. SMC-R bumper IVECO SPRI 60-100.

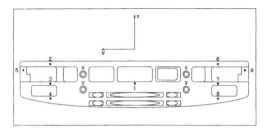

Figure 2. Frontal view of the bumper, load application (Pi) and measure points (Mj).

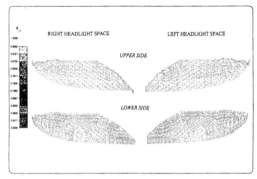

Figure 3. Distribution of the a_{11} parameter.

2 NUMERICAL ANALYSIS

2.1 Numerical model

The numerical model is a 8000 element mesh type "thin shell" with 4 and 3 nodes; the shape functions are linear with 6 dof's nodes.

This study was developed by taking into account the hypothesis of anisotropic material.

The numerical values of the physical parameters of the anisotropic material (Young's modulus E, Poisson's coefficient v) were obtained by means of the numerical simulation of the moulding process (Plastec Code).

2.2 Static analysis

The static analysis of the bumper was performed by means of a Finite Element Method code (I-DEAS VI.1).

The degrees of freedom of the nodes corresponding to each bolt were constrained in order to simulate the real clamp on an industrial truck.

The static force (1000N) was perpendicularly applied to the external surface of the bumper; the load was distributed over a circular area (diameter 70mm) (V points) (see Figure 2).

The measured quantities are the displacement components parallel to the x-axis of the bumper. For each load application point Pi (P1, P3, P5, P7, P9), the measure points Mj (M1, M2, M5, M6, M9) were chosen so as to have significant measures of displacement (load in P1, displacements in M1, M2, M5, M6, M9; load in P3, displacements in M1, M2, M3; load in P5, displacements in M1, M2, M5; load in P7, displacements in M1, M6, M7; load in P9, displacements in M1, M6, M9).

2.3 Dynamic analysis

The eigenvalues and eigenvectors of the bumper in both free-free and constraint conditions were obtained by means of the so called Simultaneous Vector Iteration (S.V.I.) algorithm of the code I-DEAS VI.1 (Normal Mode Dynamics).

Both the static and the dynamic analyses were performed by employing hardware and the I-DEAS VI.1 code kindly provided by Centro Ricerche Fiat.

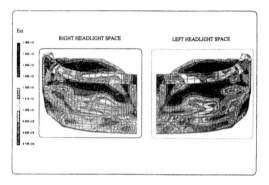

Figure 4. Distribution of the Young's modulus E_{22}.

3 RESULTS

By means of the Plastec Code simulating the moulding process, the fibre orientation and distribution and the mechanical properties of the composite material were obtained.

The fibre alignment is indicated by means of the all parameter (described in the PLASTEC Theoretical Manual) which ranges from 1 for parallel fibres to 0.5 for random fibres.

From the weight and the orientation distribution of the fibres the anisotropic mechanical properties (Young's modulus E. Poisson's coefficient v) of the material in different zones of the numerical model were evaluated. The Young's modulus was obtained both in the maximum alignment direction (E_{11}) and in the corresponding perpendicular one (E_{22}). The values predicted by the simulation for the tensile modulus (E_{11} and E_{22}) range between 9700 to 15300 MPa; the average value of the tensile modulus utilized by Bragiè et al. (1997) in the isotropic analysis was equal to 11000 MPa.

Figure 3 shows the distribution of the a_{11} parameter corresponding to both the upper and the lower side of the two (right and left) headlight spaces.

In Figure 4 is represented the Young's modulus E_{22} distribution in the whole zone corresponding to the headlight spaces (right and left).

Table 1. Static analysis results.

Measure points	Experimental results Displacement [mm]	Isotropic FEM results Displacement [mm]	Error %	Anisotropic FEM results Displacement [mm]	Error %
	Load application points				
	P1				
M1	2.99	2.63	-12.04	2.60	-13.40
M2	-0.48	-0.49	2.08	-0.48	0.00
M5	-0.80	-0.72	-10.00	-0.74	-7.50
M6	-0.40	-0.40	0.00	-0.39	-2.50
M9	-0.61	-0.59	-3.28	-0.60	-1.64
	P3				
M1	-0.34	-0.30	-11.76	-0.31	-8.82
M2	4.67	4.22	-9.64	4.64	-0.64
M3	3.43	3.47	1.17	3.44	0.29
	P5				
M1	-0.81	-0.74	-8.64	-0.77	-4.94
M2	11.25	12.64	12.36	12.36	9.78
M5	18.84	22.80	21.02	21.62	14.76
	P7				
M1	-0.31	-0.25	19.35	-0.27	-12.90
M6	3.87	3.88	0.26	3.89	0.52
M7	2.98	3.29	10.4	3.10	4.03
	P9				
M1	-0.68	-0.58	14.71	-0.67	-1.47
M6	9.30	11.4	22.58	11.21	20.54
M9	16.25	20.7	27.38	20.30	24.92

Table 2. Modal analysis results in free-free conditions.

Modes	Experimental results Frequency [Hz]	Isotropic FEM results Frequency [Hz]	Error %	Anisotropic FEM results Frequency [Hz]	Error %
1	15.09	14.11	-6.49	13.57	-10.07
2	29.24	24.82	-15.12	24.43	-16.45
3	43.31	38.81	-10.39	39.11	-9.70
4	55.33	43.56	-21.27	44.05	-20.39
5	59.33	54.29	-8.49	55.41	-6.61
6	69.53	59.78	-14.02	59.95	-13.78

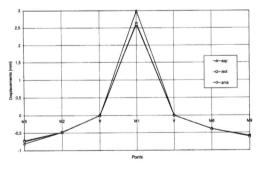

Figure 5. Displacement values corresponding to the load application point P1.

In Table 1 for each load application point the displacements of the measure points are represented: in the first column the experimental results obtained by Curà et al. (1996); in the second column the numerical ones obtained by Bragiè et al. in a numerical analysis performed by means of the isotropic FEM model and in the third column the numerical results obtained in the present work by means of the anisotropic model.

For both the isotropic and anisotropic numerical results the errors with respect to the experimental ones are calculated.

In Figure 5, for example, the displacements corresponding to the load application point P1 are shown.

Table 2 shows experimental and numerical (both isotropic and anisotropic) values of the natural frequen-

Table 3. Modal analysis results in constrained conditions.

Modes	Experimental results [C=30Nm] Frequency [Hz]	Isotropic FEM results Frequency [Hz]	Error %	Anisotropic FEM results Frequency [Hz]	Error %
1	28.86	23.92	-17.12	23.73	-17.77
1bis	-	25.22	-	24.87	-
2	38.13	36.64	-3.91	36.98	-3.02
3	46.16	45.34	-1.78	44.63	-3.31
4	56.94	51.88	-8.89	51.76	-9.10
5	63.07	55.22	-12.45	56.54	-10.35
6	72.28	64.38	-10.93	67.07	-7.21

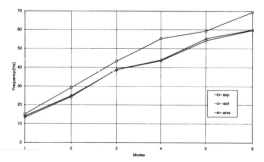

Figure 6. Natural frequency values in free-free conditions.

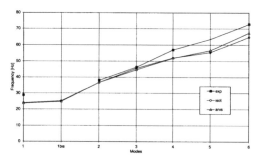

Figure 7. Natural frequency values in constraint conditions.

cies of the bumper in free-free conditions. Experimental values were obtained by Curti et al. (1996), numerical isotropic values were obtained by Bragiè et al. and anisotropic values are calculated in the present work.

Table 3 shows experimental and numerical (both isotropic and anisotropic) values of the natural frequencies of the bumper in constrained conditions.

Also in this case experimental values were obtained by Curti et al. (1996), numerical isotropic values were obtained by Bragiè et al. and anisotropic values are calculated in the present work.

For both the isotropic and anisotropic numerical dynamic analyses in free-free and in clamped conditions the errors with respect to the experimental values are calculated.

In Figures 6 and 7 the experimental and numerical values of the natural frequencies respectively in free-free (Table 2) and clamped conditions (Table 3) for varying the order of the mode are presented.

4 CONCLUSIONS

Reasonable agreement between experimental and numerical FEM results (both isotropic and anisotropic) is shown in Tables 1, 2 and 3; the numerical isotropic analysis was performed by means of an appropriate choice of physical and geometrical parameters (Young's modulus, Poisson's coefficient, thickness); the anisotropic results were obtained by utilizing mechanical properties predicted by means of a numerical simulation of the moulding process (Plastec Code).

Numerical displacements (Table 1) both obtained from the isotropic and anisotropic static analysis are in a good agreement with the experimental ones, even if the numerical models seem a little bit stiffer.

In the dynamic analysis both isotropic and anisotropic models provide natural frequency values which are practically equal.

Numerical frequencies both calculated in free-free and clamped conditions are in a very good agreement too with the corresponding experimental ones, even if experimental values are a little bit higher.

Finally it can be pointed out that Plastec Code simulation provides a satisfactory predicted anisotropic model. This moulding process simulation method can be considered a powerful tool in both static and dynamic analysis.

There is also a very good agreement between isotropic and anisotropic numerical results.

In general, it is possible that FEM codes can be used for complex shaped composite structures already in the design phase.

Nevertheless experimental tests performed on SMC plates and prototypes are necessary to determine the average isotropic mechanical properties to use in the numerical analysis.

Experimental tests are not required if the ability to simulate the filling of the mould together with the final fibre orientation allow reliable determination of the anisotropic characteristics of the component.

This study has shown that the simulation moulding process (Code Plastec) can also be used to predict the anisotropic properties of a composite material in order to obtain the necessary average isotropic values to introduce in a less complex and expensive corresponding isotropic FEM model.

REFERENCES

Bragiè D., Curà F., Curti G., Indino E. 1997. Metodologie avanzate di analisi numerica di componenti in materiale composito per veicoli industriali. *Proceedings of TECNITEX Conference, Turin, Italy,* 19-21 november.

Curà F., Curti G. 1996. Dynamic behaviour of an SMC bumper for industrial truck. *Proceedings of the Third Seminar on Experimental Techniques and Design in Composite Material, Cagliari, Italy,* 30-31 october: vol.144, 163-178.

Curà F., Curti G., Ghirardo C., 1996. Analisi numerica e sperimentale del comportamento statico di un paraurti per veicolo industriale in materiale composito con fibre di vetro. *Proceedings of TECNITEX Conference, Turin, Italy,* 21-23 november: 55-65.

Lazzarin P., Molina G., Molinari L., Quaresimin M., 1996. Numerical Simulation of SMC Component Moulding *Proceedings of the Third Seminar on Experimental Techniques and Design in Composite Material, Cagliari, Italy,* 30-31 october: vol. 144, 191-200.

Plastec Theoretical Manual, 1991, *Technalysis 7120, Indianapolis-USA.*

This study was carried out with MURST funding of 40%.

Experimental Techniques and Design In Composite Materials 4, Found (Ed.)
© 2002 Swets & Zeitlinger, Lisse, ISBN 90 5809 370 0

Structural optimisation of an Olympic kayak

N. Petrone
Department of Mechanical Engineering, University of Padova, Italy

M. Quaresimin
Department of Management and Engineering, University of Padova, Italy

ABSTRACT: Recent improvements in Olympic kayaking are being achieved through the study of new paddling techniques and by parallel developments of paddling devices and kayak structures. In fast kayaking, the pitching behaviour of the vessel is reported by the athletes to reduce the overall performances; this effect mainly being related to the kayak bending stiffness, a structural analysis of a composite Olympic kayak was performed to increase the bending stiffness of the vessel. The complex geometry of the lower shell and the upper cover were measured and a finite element model of the kayak was developed. The mechanical properties of the different laminates present in the composite structure were experimentally evaluated and the FE model was calibrated with experimental results from bending tests. Several technical solutions for the kayak manufacturing were compared to increase the overall stiffness, giving a final improvement higher than 20% in terms of specific stiffness. The method may be transferred to other longer boats, K2 or K4, where the load synchronism and the pitching effects are even more important.

1 INTRODUCTION

The improvement of the sport performances in the Olympic kayak discipline can be achieved by the contribution of several factors, related both to the athlete's performance level and technique and to the sport devices optimisation (Kearney et al. 1979, Kendal et al. 1992). The two devices used in the kayak discipline are the paddle and the kayak vessel and the different modifications of these tools, introduced since the early beginnings of the discipline, show the positive effects of the design evolution.

Since the first introduction of the discipline at the 1963 Olympic games, several modifications to the vessel geometry and the adopted materials were introduced. The present construction dates basically from 1965 and contains some typical features that differ considerably from the early rounded shaped kayak based on the Eskimos boats. The V-shape of the lower vessel, the lifted prow and stem and the asymmetric rhomboidal shape with the maximum width shifted to the back of the paddler are the most important properties that make the kayak very fast, stable to yawing, but also unstable to rolling. Furthermore, the progressive substitution of wood with composite materials allowed an additional increment of the stiffness properties and a parallel reduction of the weight, even if the design approach to the material selection and application seems to be still related more to the craftsmen's skill and fantasy than to an engineering analysis.

The structural behaviour of a 5,2 m long Kl vessel

results are of interest in terms of stiffness properties that may be related to the pitching phenomena in the water, reported by the athletes to be detrimental to the final results. This aspect is even more important when considering the multiple K2 and K4 boats, respectively 6,5 and 11 m long, where the longer vessel structures have to sustain the forces applied by the athletes, often not acting synchronously.

The knowledge of the loads applied by the athlete to the kayak is still a research topic: an acquisition device has been recently developed for measuring the forces applied by the paddler to the kayak vessel at the two body contact points, the saddle and the footpad (Petrone et al. 1998a) and only the first results are available (Petrone et al. 1998b). As reported also by the athletes, the typical action adopted in the modern paddling technique is an intense foot push synchronous with the arm pulling the paddle in the water at the same side of the kayak: therefore the vessel receives forces of opposite sign at the footpad and the saddle, with a possible bending action in the longitudinal vessel plane. Together with the pitching on the water, this bending effect was seen to have strong influence on the structural behaviour and therefore on the kayak performances and was then investigated in the present study. A comparative experimental set-up was defined to evaluate the longitudinal bending behaviour of a high quality Olympic composite kayak and a numerical model was developed to analyse via FE simulations the possible improved solutions for a better performing kayak.

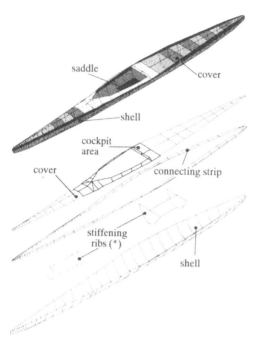

Figure 1. An Olympic kayak and the main vessel parts (*ribs not usually present).

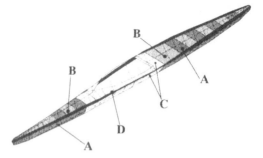

Figure 2. Layout of the materials used in the kayak vessel manufacturing.

2 THE OLYMPIC KAYAK

A general view of an Olympic K1 kayak is reported in Figure 1, where the main parts of the vessel are also indicated.

The kayak is built by connecting two main parts, the lower shell and the upper cover: a U shaped hole in the cover, named cockpit, allows the athlete to sit on the saddle and gives to the boat its typical aspect. The area surrounding the cockpit has been highlighted because of the reinforced structure there applied due to the presence of the hole. The structures internal to the kayak that are also indicated in Figure 1, such as stiffening ribs and bulkheads, are not present in the actual boat, but were introduced as possible optimised solutions.

The present regulations for the Olympic games allow for K1 boats 5,2 m long and 0,51 m wide, with a minimum mass of 12 kg without the athlete: it is interesting to note that the minimum mass is usually reached by application of an additional mass under the saddle, because the kayak is, normally, lighter than 12 kg.

The modern kayaks are normally made in composite materials which have substituted the more traditional wooden structures. The vessel under analysis was built by using laminate and sandwich structures; three types of dry fabric were used as raw materials for the laminates: carbon fabric (CF), kevlar fabric (KF) and hybrid kevlar-carbon fabric (KCF); the same fabrics and a Firet Coremat XM core (combination of non-woven and expanded plastic micro-spheres) (XM) were used to produce the sandwich structures. During the laminate and sandwich preparation, the dry fabrics were hand-laid up in the vessel moulds and manually impregnated with epoxy resin; the fabric were laid up with their warp direction parallel to the longitudinal axis (Z) of the kayak. The vacuum bagging technique (without autoclave) and infrared lamps were used for the vessel manufacturing. After separate moulding, the cover and the shell are connected by an internal strip of kevlar tape laminated over the local structure of the two parts of the vessel.

The different laminates and sandwiches present in the vessel were grouped into 4 different materials as described in Table 1 and their particular arrangement in the different areas of the kayak is presented in Figure 2. It should be noted that the shell sandwich structure is reinforced in correspondence to the cockpit and that the cover basic laminate is stiffened with several sandwich strips in the front and rear part; the cover cockpit area also presents a reinforced sandwich structure.

Table 1. Layout and description of the materials used in the kayak vessel manufacturing.

Material	Areas of application	Structure	Lay-up Inside ⇨ Outside	Laminate thickness (mm)
A	cover and shell	sandwich	KF/XM/KCF	0.75/1.8/0.75
B	cover	laminate	KF/KCF	0.75/0.75
C	cockpit and shell (below cockpit)	sandwich	KF/CF/XM/KCF	0.75/0.75/1.8/0.75
D	connecting strip	sandwich	KF/CF/XM/KCF	0.75/0.75/1.8/0.75

3 ANALYSIS METHOD

According to previous structural analysis on composite components (Lazzarin et al. 1995), the procedure adopted for the analysis and optimisation of the kayak bending behaviour can be described as a sequence of different steps:

- geometrical 3D CAD modelling by reverse engineering on the kayak vessel;
- FE modelling of the kayak vessel;
- experimental evaluation of the mechanical properties of the materials used in the kayak manufacturing;
- experimental evaluation of the kayak bending stiffness and calibration of the FE model;
- FE optimisation.

The actual geometry of a K1 kayak vessel was measured, due to the lack of drawing documentation; a 3D model was then generated by means of a CAD software. The geometry was imported into a Finite Element code for the development of a suitable numerical model of the kayak and the simulation of an experimental set-up for comparison and calibration of the model.

The composite material properties were experimentally evaluated by extraction of standard specimens from the same laminates used in the kayak manufacturing, due to the lack of literature referring to the particular technologies adopted.

In parallel, a meaningful comparative experimental test for the longitudinal bending analysis was identified, with simple loading and restraint conditions, to allow for comparison of kayaks from different manufacturers and for a correct calibration of the numerical model.

Finally the calibrated numerical model was used for the simulation of the bending behaviour of different optimised solutions aiming to increase as much as possible the absolute and specific stiffness.

4 DEVELOPMENT OF THE KAYAK NUMERICAL MODEL

The first part of the study consisted in the manual measuring of the kayak geometry that could be based either on the vessel or on the moulds available at the manufacturer's workshop. This can be seen as a problem of reverse engineering needed when, as in this case, the production of a component is not based on drawings but on established craftsmen skills and tools.

The frame of reference adopted for the boat was a XYZ system with the Z axis parallel to the longitudinal vessel direction and the Y axis vertical with respect to the water, placed at the front tip of the boat.

To reach an higher precision, the external vessel surface was measured on the two available moulds, used for the shell and the cover, instead of making reference to the moulded kayak vessel. The distance of any section was taken from the tip of the boat along the Z axis and the ZY plane was assumed to be a symmetry plane. Then, by means of a special tool equipped with two

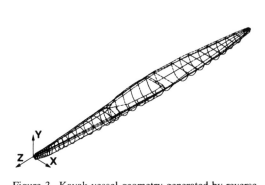

Figure 3. Kayak vessel geometry generated by reverse engineering.

rulers acting in the X and Y directions, 34 transverse sections of the boat were measured on each of the two moulds of shell and cover. At each section with known Z value, 10 points on the shell surface and 5 points on the cover surface were taken. The sequence of points describing each measured section were input in the 3D software INTERGRAPH Microstation and connected as transverse sections of the boat: the obtained external geometry, shown in Figure 3, was then imported in ANSYS 5.4 FE code via an IGES file. Due to some inconsistencies present in the IGES formats, the transverse sections were redrawn with b-splines and the shell and cover surfaces were defined between consecutive sections within the ANSYS pre-processor.

The composite external and internal surface of the kayak vessel were modelled by using SHELL 91 layered orthotropic elements while SHELL 93 isotropic elements were used for the plate simulating the saddle where the athlete is sitting.

The footpad, the rudder rod and rudder hole were not modelled, considering their low contribution to the longitudinal bending behaviour.

The aim of the analysis being the investigation of the kayak bending stiffness, the mesh density was chosen close enough to correctly describe the bending behaviour but without small detail modelling that a local stress field analysis would have required; this allowed the model size and the solution time to be kept reasonably low.

The resulting final FE model of the kayak, shown in Figure 4, consisted of 3200 parabolic shell elements for a total number of 9400 nodes and 56000 active DOFs; the size of the FE model file (FILE.DB) resulted approximately of 20 Mb, the result file (FILE.RST) of 20 Mb and the temporary files required about 320 Mb with the Frontal solver. The CPU time needed for a linear elastic solution was about 10 minutes on a SUN Ultra 60 workstation.

As previously pointed out the model was developed for a stiffness analysis; therefore it has to be remeshed for a reliable stress analysis that could be performed once the loads applied to the kayak during the paddling action are available.

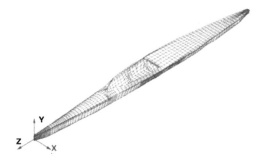

Figure 4. The kayak FE model.

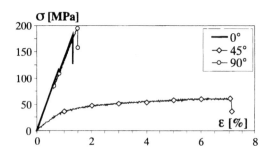

Figure 5. Stress-strain curves for the hybrid kevlar-carbon fabric laminate (KCF).

5 MATERIAL PROPERTIES EVALUATION

The kayak vessel is mainly made of composite laminates and sandwich structures obtained by manual fabric impregnation, hand lay-up and cold curing vacuum bagging. This manufacturing technology allows the production of good quality kayaks at acceptable costs but it can have a strong influence on the laminate properties due to the reduced level of control of both resin content and curing process. It is therefore very important for a reliable numerical simulation of the kayak stiffness behaviour to know the actual properties of the laminates. This was obtained by an accurate experimental evaluation on specimens produced with the same process used in the kayak manufacturing. As mentioned before, three types of dry fabric were used as raw materials for the laminate and sandwich structures: plain weave balanced carbon fabric (175 g/m^2), plain weave balanced kevlar fabric (175 g/m^2) and plain weave hybrid kevlar-carbon fabric (175 g/m^2), all manually impregnated with polyester resin.

The static properties of the different materials were evaluated on standard ASTM 3039 specimens, instrumented with HBM 6/120LY13 strain gauges. For each material, specimens oriented at 0°, 45° and 90° with respect to the fabric warp direction were tested, in order to otain the four independent elastic constants (E_L, E_T, ν_{LT} and G_{LT}) describing the in-plane orthotropic behaviour of the laminates and required for the FE simulation.

The shear modulus of the Firet Coremat XM core was evaluated by using a simplified relationship proposed in (Caprino & Teti 1989) on the basis of the results of a set of three point bending tests carried out on sandwich specimens.

The results of the static characterisation are summarised in Table 2 and the stress-strain curves for the hybrid kevlar-carbon laminate are plotted, as an example, in Figure 5.

It is possible to observe the relatively low values obtained for the laminate elastic and strength properties if compared to those of laminates made of the same material moulded in an autoclave. This can be explained by the high resin content related to the manual impregnation of the dry fabrics and the cold-cure manufacturing.

6 EXPERIMENTAL TESTS ON THE KAYAK

The bending stiffness of the kayak was experimentally evaluated after the definition of a suitable test set-up, intentionally representative of the in-service loading conditions related to the pitching and longitudinal bending behaviour. The kayak was mounted on a universal test bench, simply supported by two steel strips, in isostatic restraint conditions, and loaded with calibrated weights applied to the saddle, simulating the presence of the paddler. The restraint position along the kayak longitudinal axis was chosen in order to obtain a stiffness parameter representative of the contribution of the complete kayak structure, which is almost immersed in the water during the paddling, being the floating line quite close to the connection strip between the cover and shell.

Figure 6 presents a global view of the kayak mounted on the testing bench and Figure 7 shows a detail of the

Table 2. Strength and elastic properties of the materials used in the kayak vessel manufacturing.

Material	Tensile strength [MPa]	Strain to failure [%]	Tensile modulus [MPa]	ν_{LT}	G_{LT} [MPa]
Carbon fabric (CF) 0°	197.0	1.14	28930 ($E_L=E_T$)	0.188	1640
Kevlar fabric (KF) 0°	176.2	2.33	10030 ($E_L=E_T$)	0.157	1030
Kevlar-Carbon fabric (KCF) 0°	179.7	1.32	14680 E_L	0.180	980
Kevlar-Carbon fabric (KCF) 90°	197.7	1.50	13990 E_T		
core (Firet Coremat XM)	-	-	-	-	Shear modulus 50

Figure 6. Experimental test set-up for the bending test on the kayak.

Figure 7. Prow restraint and dial gauges for the evaluation of the cover and shell displacements.

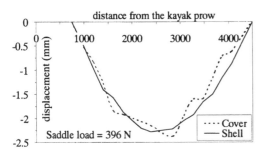

Figure 8. Comparison of the experimental cover and shell deformed shapes.

prow restraint and the dial gauges used for the evaluation of the local displacement along the kayak due to the load application.

The stiffness test consisted in the incremental application of calibrated weights (up to ≈ 400 N) to the kayak saddle and in the measurement of the resulting displacements of both the cover and shell in 14 sections along the longitudinal kayak plane (YZ). The measurements, once corrected to account for the unavoidable contribution due to the restraint compliance, allowed for the definition of the overall deformed shape which indicated the influence of the different vessel zones and the areas to be modified or improve during the optimisation phase.

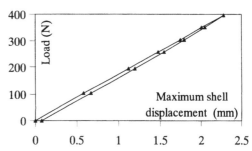

Figure 9. Load vs. displacement plot for a complete bending test.

Figure 8 shows a comparison of the displacements measured on the kayak cover and shell, between the constraints, during a bending test: it is evident the non uniform deformed shape of the cover which is mainly due to the presence of the cockpit hole.

On the basis of the experimental results and considering the more uniform deformed shape of the shell, a definition for the kayak bending stiffness was introduced as the ratio between the applied load and the relevant maximum displacement of the shell. A typical plot of the applied load vs. the maximum shell displacement is shown in Figure 9. The kayak stiffness was evaluated by a linear best fit of the experimental results of at least three repeated tests.

7 FE MODEL CALIBRATION

The experimental analysis of the kayak bending behaviour proved to be fundamental for the definition of a reliable FE model. The comparison between experimental and numerical results, in terms of deformed shape of the kayak under load, allowed the calibration of the kayak FE model, by choosing the numerical restraints configuration suitable to better represent the experimental conditions.

Figure 10 shows the load and restraint conditions applied to the kayak FE model for the simulation of the bending test. The calibrated weights applied to the kayak saddle were simulated as a concentrated force and the displacements of the nodes placed in the sections supported with the steel strips were restrained in order to obtain an isostatic support configuration.

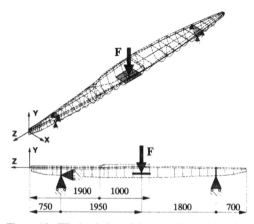

Figure 10. FE simulation of the experimental bending test (load and restraints positions).

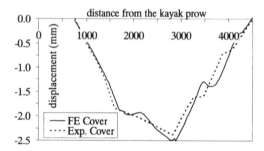

Figure 11. Comparison between experimental and FE cover deformed shape.

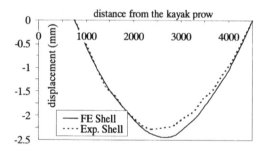

Figure 12. Comparison between experimental and FE shell deformed shape.

Figures 11 and 12 compare the experimental and numerical kayak deformed shapes for the cover and shell respectively, obtained by analysing the kayak behaviour with the calibrated FE model. The maximum difference between numerical and experimental results was less than 10%, which is acceptable considering that

Table 3. Description of models analysed during the optimisation phase.

Model	Properties and modifications
1	Reference model (as described in table 1)
2	Front and rear transversal ribs
3	Front and rear longitudinal bulkheads
4	Front and rear, longitudinal bulkheads and transversal ribs
5	Additional carbon lamina on the connecting strip between cover and shell
6	Model 4 and model 5 combined
7	Model 5 and additional carbon lamina on the cover cockpit area
8	Cover and shell cockpit areas: substitution of the external kevlar-carbon lamina with a carbon lamina; remaining shell areas: substitution of the external kevlar-carbon lamina with a kevlar lamina
9	Substitution of the external kevlar-carbon lamina with a carbon lamina in all the sandwich areas

the aim of the analyses was a comparative optimisation of the bending stiffness. On the other hand, the maximum deflection was evaluated with an error of less than 5%.

8 OPTIMISATION ANALYSIS AND RESULTS

The increase in the absolue and specific stiffness properties of the kayak were considered the main aims of the optimisation phase. However, during this phase, some limitations had to be taken into account: in particular the already mentioned minimum kayak weight of 12 kg imposed by the present regulations for the Olympic games together with a defined geometry, the technical feasibility of the possible design solutions for the kayak and, last, the production cost of these solutions.

By keeping in mind these constraints, the optimisation of the kayak performances was accomplished by FE simulation of the bending behaviour of eight design solutions for the kayak, including the modification of the material lay-out and lay-up and also the introduction of stiffening elements. The different solutions analysed during the optimisation phase are listed in Table 3, where the modifications introduced with respect to the existing kayak configuration (reference model) are presented.

With particular reference to models 2, 3, 4 and 6, the introduction of stiffening ribs was considered a possible solution to increase the stiffness, the position and shape of these stiffening elements are shown in Figure 13.

Table 4 presents the mass and the stiffness evaluated for each model during the optimisation phase; the absolute and specific properties variations with respect to the original model, taken as reference, are also reported in Table 4 and shown in Figure 14.

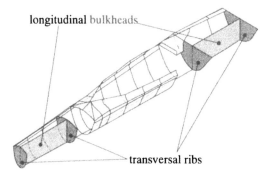

longitudinal bulkheads

transversal ribs

Figure 13. Position of the stiffening ribs inside the kayak vessel.

Table 4. Mass and stiffness properties of the different kayak models.

Model	m (kg)	K (N/mm)	Ks (N/mmkg)	Δm (%)	ΔK (%)	ΔKs (%)
1	10.3	172.3	16.7	0	0.0	0.0
2	10.7	173.4	16.2	3.9	0.6	-3.2
3	11.1	177.7	16.0	7.8	3.1	-4.3
4	11.5	178.9	15.6	11.7	3.8	-7.0
5	11.0	185.3	16.8	6.8	7.5	0.7
6	12.0	193.1	16.1	16.5	12.0	-3.8
7	11.3	189.4	16.8	9.7	9.9	0.2
8	10.6	187.0	17.6	2.9	8.5	5.4
9	10.7	217.7	20.3	3.9	26.3	21.6

(m: estimated mass, K: bending stiffness, Ks: specific bending stiffness, Δm: mass variation, ΔK: bending stiffness variation, ΔKs: specific bending stiffness variation)

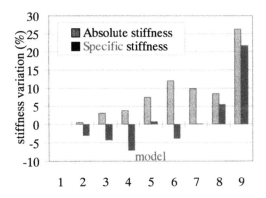

Figure 14. Variation in the absolute and specific bending stiffness for the different kayak models with respect to the reference model.

The first consideration to note is the limited improvement in the absolute stiffness due to the introduction of both transversal and longitudinal stiffening ribs; the contribution is even negative if the specific stiffness is considered. On the other hand the absolute and specific stiffness properties of the kayak turn out to be very sensitive to the presence of carbon fibre reinforced laminae, as could be expected due to their higher tensile modulus.

The results of model 5 also highlight the influence of the connecting strip between the shell and cover on the overall kayak behaviour.

The best results in terms of both absolute and specific properties are those related to model 9, built with an extensive use of carbon fibre reinforced laminae.

However, it is important to remember that the use of carbon laminae could reduce the impact resistance of the vessel laminates, which is another important feature of a kayak vessel. In fact during some long races, like the 10000 metres, the collisions among two or more kayaks are quite frequent with possible serious damage to the vessels.

A possible way to increase the impact resistance of the kayak vessel without increasing the weight is the use of kevlar laminae and therefore a more reliable, although less stiff solution can be represented by model 8 where the use of carbon fibre reinforced laminae is limited to the central zone of the vessel.

Finally, it has to be pointed out that, although an increase of the kayak weight is always associated to the alternative solutions represented by the analysed models, the minimum weight of 12 kg has never been exceeded.

9 CONCLUSIONS

The bending behaviour of a composite Olympic kayak has been investigated by means of a FE model developed in ANSYS environment and bending experimental tests on the kayak.

The bending stiffness of several design solutions for the kayak manufacturing has been compared: and a general stiffness increase, up to a maximum of 25%, has been found for all the proposed solutions.

Only an extensive use of carbon fibre reinforced laminae allows, however, a strong increase of the specific bending stiffness, although this could reduce the impact resistance of the vessel laminate.

The FE model and the bending test device developed during this work could also prove useful for the stress analysis of the kayak once the loads due the paddling action are available.

REFERENCES

Caprino, G. & R. Teti 1989. *Sandwich Structures Handbook*, Padova, Italy: Il Prato.
Kearney, J.T., Klein L. & Mann R. 1979. The elements of style: an analysis of the flatwater canoe and kayak stroke. Canoe, 7(3): 18-20.

Kendal, S.J. & Sanders R.H. 1992. The technique of elite flatwater kayak paddlers using the wing paddle. International Journal of Sport Biomechanics. 8: 233-250.

Lazzarin P., Meneghetti G., Petrone N. & Quaresimin M. 1995. Structural Analysis of a Racing Motorcycle Composite Swing Arm. In M.S. Found (ed) *Experimental Techniques and Design in Composite Materials - Proc of 2nd Int. Seminar, Sheffield September 1994*, 350-364. Sheffield: Sheffield Academic Press.

Petrone N., Quaresimin M. & Spina S. 1998. A load acquisition device for the paddling action on Olympic kayak. In Experimental Mechanics, Advances in Allison (ed.) *Design, Testing and Analysis - Proc. of 11th ICEM, Oxford, 24-28 August 1998*, Vol.2 817-822. Rotterdam: Balkema.

Petrone N., Quaresimin M. & Spina S. 1998. Acquisition and analysis of the paddling loads on Olympic kayak. in Proceedings of XXVII AIAS National Conference, Perugia (Italy) 9-11 September 1998 Vol. II, pp. 1013-1022. (in Italian).

Experimental Techniques and Design In Composite Materials 4, Found (Ed.)

Design of a composite stabiliser bar

M.S. Found & C. de Luis Romero
Department of Mechanical Engineering, University of Sheffield, UK

ABSTRACT: A conceptual analysis for the design of a stabiliser bar using fibre reinforced materials is described. The critical design aspect is the stabiliser equivalent stiffness. An optimum filament winding parameter's combination is determined for a composite/aluminium hybrid stabiliser bar that is expected to perform in a similar manner as the steel stabiliser bar it replaces. A comparison between both stabiliser bars' performance is carried out detailing the advantages and disadvantages of the new design. Finally, an optimisation analysis is done for the bonded joint between the composite tube and the aluminium stabiliser legs.

1 INTRODUCTION

Since the advent of the motor vehicle, there has been a drive for a continuous reduction of the overall weight which has now become more important. A reduction of the weight will imply a reduction of the fuel consumption, a commodity becoming more and more scarce.

There has been continual development programmes in order to keep the weight of the different components to a minimum without affecting the mechanical properties and performance of the vehicle. Alternative materials to replace the classical steel have been studied, many of the elements that are not supporting excessive load have been manufactured in plastic materials. Nevertheless, the components that perform any mechanical function take a high partial weight of the vehicle and more resistant materials need to be used. Among them, fibre reinforced composites are gaining a greater acceptance due to both their low density and good mechanical properties.

The designing of vehicles' structural components requires a detailed analysis, that becomes more complex when anisotropic materials are introduced. Larger factors of safety are often required when designing with relative unknown materials, followed by experimental analysis of the whole component. These and other reasons contribute to make the introduction of fibre reinforced materials in the automotive industry very slow for components subjected to some important loading conditions.

Designs of structural components in fibre reinforced composites have been performed on suspensions (de Goncourt 1987) and on transmission shafts (Belingardi 1990). The aim of the present paper is to point out a design process route when designing a replacement stabiliser bar for an isotropic material, using new materials

that offer higher potential weight reduction together with better mechanical properties. The use of long fibre reinforced plastics, provides the designer with the possibility of *tailoring* the material properties, designing a material that accommodates to the applied stress in different directions along the component. An optimum material use is then obtained through an engineering analysis. Also, importantly for these materials, a manufacturing route is considered at the initial design stage.

1.1 *The concept*

A stabiliser bar is fitted into a vehicle suspension in order to increase the stiffness under a roll movement, therefore imposing an additional resistance to the roll movement of the car during manoeuvring.

Latest configurations consist of angled torsion rods mounted on the body in rubber bushes so that they can rotate, the bent ends are connected to the steering heads and suspension arms. Modern vehicles incorporate an independent front suspension to improve the ride performance in such a way that softer springs may be used for normal vertical loads transmitted from the wheels. However, when the vehicle is cornering, the action of

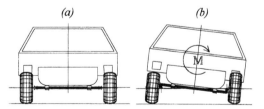

Figure 1. Stabiliser bar function a) Inactive b) Working under body roll.

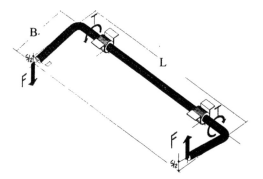

Figure 2. Stabiliser bar solicitations.

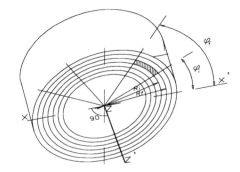

Figure 3. Equivalent moduli calculation.

the stabiliser bar will provide the necessary stiffness resistance to oppose body roll. The final shape usually is a "U", although geometry details depend on other requirements due to the allocations of additional vehicle components.

The stabiliser bar transmits a torque from the inside wheel, which is subjected to less load, to the outside wheel which is now more heavily loaded. If both right and left wheels are moving up or down together at the same rate, such as would happen when driving over a speed bump, the bar remains untwisted, it simply rotates and provides no resistance to the vertical wheel motion. These characteristic allow a car to be fitted with relatively soft springs to absorb road irregularities, while the stabiliser bar keeps roll under control.

1.2 Stabiliser equivalent stiffness

The vertical displacement between the free ends is a combination of the twisting angle of the bar and the deflection of the legs. The equivalent bar stiffness is understood to be the applied force divided by the relative displacement.

Neglecting the bending contribution, a traditional equation for the stiffness is given as:

$$K_x = \frac{F}{X} = 2\frac{G \cdot J}{L \cdot B^2} \tag{1}$$

Where:

G	Shear modulus
J	Second polar moment of area

The equivalent bar stiffness is the principal design parameter. A stiff bar means a smaller displacement between the free ends, or a harder response. A compliant bar will imply an excessive roll movement, uncomfortable for commercial vehicles and inappropriate for high performance racing cars.

1.3 Analysis procedure

The replacement composite bar is designed to produce

the same stiffness as the steel bar, as determined by the previous designers.

The most suitable composite manufacturing technique for this component is filament winding. The bar can be assumed to be formed by different cylindrical layers of a very small thickness and hence the laminate theory can be applied for every single layer. The mechanical properties of a fibre reinforced material are determined by the principal parameters: fibre and matrix properties, winding angle and fibre volume fraction. These parameters, together with the bar dimensions, can be combined to obtain the required stiffness at the lowest weight and material utilisation.

2 BASIC EQUATIONS

The actual steel bar is designed for stiffness. Evaluation using both analytical and experimental stress analysis shows that the steel bar is overdesigned in terms of strength. Hence, the new bar is also designed for stiffness and then examined for strength in order to ensure that the design is safe and to determine appropriate factors of safety.

2.1 Cross section equivalent stiffness

A design using isotropic materials is based in the longitudinal stiffness (Young's modulus E) and the shear modulus (G). In the same way, the stiffness for a composite tube can be determined for both the normal (E·I) and shear (G·J) modes.

In doing so, the cross section is evaluated, for a general case, in circular sectors of different stiffness moduli (E & G), these values depend on the composite parameters and the geometric values (I & J) that depend on the dimensions.

The section is assumed to be linearly deformed as a consequence of the displacement compatibility equations and the strain in the section is given by:

$$\varepsilon_i = \frac{M}{(EI)_{eq}} r_i \tag{2}$$

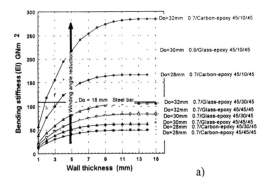

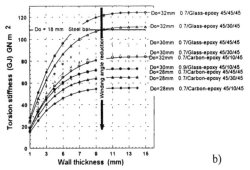

Figure 4. Cross section equivalent strain moduli.
a) Bending moduli b) Torsion moduli.

Where: $EI_{eq} = \sum EI_i$ (for every i circular sector).

r_i defined in Figure 3.

The stresses in every sector can now be calculated using traditional laws of physics. The case for torsion is done in the same way, using the torsional stiffness $(GJ)_{eq}$.

The fibre orientation angle determines to a great extent these moduli values. Very small angles are more suitable for bending resistance, meanwhile a 45 degrees winding angle is favourable for torsion loading. The component under study presents a combined bending-torsion loading system and hence an adequate winding angle has to be determined for a high stiffness bar.

In Figure 4, is represented the variations of the section stiffness for different winding angle combinations in different layers. The effect described in the paragraph above can clearly be observed in these results. With a wound material it is possible to play with the bending and torsion contributions, both moving in opposite directions.

Winding machines can vary the winding angle along the component to adapt to the strength requirements, nevertheless this produces continuous overlaps or gaps and the effective variation length depends on other mechanical parameters.

A sensitivity analysis has also to be done by changing the angle along the bar. Although the stress levels can be easily reduced in the highest bending loaded locations, the stiffness of the whole bar is not significantly varied from a maximum value for 45 degrees. The reason for this is that the contribution of the torsion resistance to the total bar's stiffness is about 80% and hence an aim has to be to make the bar stiffer in the shear direction.

3 OPTIMISATION PROCEDURE

In order to obtain the best combination of parameters for the stiffest bar, using the smallest amount of material, the winding angle has been changed for a given volume fraction. The bar is assumed to be formed by a central hollow composite tube and two aluminium legs.

3.1 Material selection

The filament and matrix mechanical properties (the stiffness moduli in this particular case) are parameters of a great importance to play with. Glass fibres (E $\cong 70$GPa) are the most suitable for filament winding techniques at 45 degrees due to the low contribution of the fibre in the bulk material properties, since the matrix determines essentially the properties in a higher amount. Carbon fibres (E $\cong 200 - 500$ GPa) present a much higher modulus, lower density but are substantially more expensive.

In Figure 5, this effect can be easily noticed. When designing for bending, (small winding angle) the cross section equivalent moduli rises proportionally to the fibre moduli. In the case of torsion (45°) the relation is no further linear but increases at a very low rate to become almost constant.

Due to this, no further stiffness benefit is obtained by using better fibre properties at the high cost that it entails.

After this study it can be said that the material has almost reached it highest stiffness for a fibre moduli of about 200 - 300 GPa. The only way to increase the stiffness is to change the section dimensions if the rest of parameters remain unchanged.

3.2 Application

Engineering specifications indicate a value of the stroke of ±30 mm for testing purposes. The fatigue life for this value has to be over a given value. This displacement is the value that has been taken as a reference for the case of the hybrid stabiliser bar replacement. The load applied at the free ends to reach this displacement level is of 2160 N for the steel bar.

The central bar is subjected to a torque of 261.6 N·m and a maximum bending moment of 252.8 N·m, meanwhile the aluminium leg presents a 260.7 N·m bending at the joint and a negligible torque for a 2160 N applied load.

Although the design of the composite tube is done to stiffness, this does not happen when designing the alu-

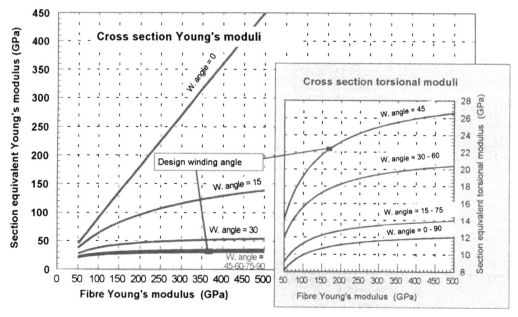

Figure 5. Section equivalent modulus versus fibre modulus.

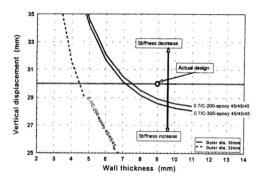

Figure 6. Stiffness design map.

minium legs for which the stiffness is not the limiting parameter but the strength resistance. The stiffness contribution of the legs is due uniquely to bending and can be ignored without an important error.

In the following graph the displacements at the free ends are represented for different combination of the composite parameters and dimensions. As the same stiffness level is desired, the design points are those for S/2 = 30 mm for an applied load of 2160 N.

Among all the possible points on the design map shown in Figure 6, one must be selected that represents the most suitable combination of section diameter and wall thickness for the given fibre properties.

4 REPLACEMENT BAR

After a detailed sensitivity analysis, it has been concluded that the optimum winding angle when designing for stiffness is 45 degrees constant along the bar. A smaller angle in some of the layers along the most severely bending loaded region will reduce the bending deflection but increase the twisting angle, the contribution of bending is less than 25% of the torsion and hence the combined effect is unfavourable. A medium modulus carbon fibre has been finally selected as a compromise with the geometrical dimensions. Although glass fibres are much cheaper, its use leads to an excessive diameter, not suitable for this application. The final diameter has to be substantially larger than that of the original steel stabiliser bar, but this is an inevitable situation if no other dimensions are changed. An important reduction in the overall weight is, nevertheless, achieved by making a hollow bar of a lighter material. The finally selected materials are indicated in Table 1.

4.1 Dimensions

The diameter of the bar is of 30 mm with a wall thickness of 7 mm to obtain the same equivalent stiffness as that for the steel bar. The aluminium legs are designed to fit at the tube ends and to resist the bending and torsion stresses.

The hybrid bar has been fully designed for stiffness and the resulting dimensions are far higher than that required for strength considerations. The factor of safety

Table 1. Material selection for the hybrid stabiliser bar

Material	E GPa	S_u MPa	S_f MPa	ρ kg/m^3
360/380 Die cast Aluminium alloy	70	300	150	2650
Matrix				
epoxy	4.5	35	-	1100
Reinforcement				
Carbon fibre	250	3400	-	1750
Composite				
$V_f = 0.7/\alpha = 45$	16.5	2675*	1500	1555

* In principal directions

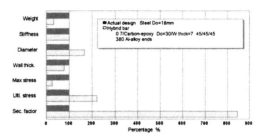

Figure 8. Comparison of parameters for hybrid and steel bars.

Figure 7. Hybrid bar geometry.

Table 2. Comparison of stresses

	Max. Stress MPa	Ultimate stress MPa	Safety factor	Fatigue life Cycles
Steel bar	625.8	1100	1.76	>10^6
Hybrid bar				
Aluminium leg	140	305	2.2	>10^8
Composite tube	150	2675	17.8	>10^8

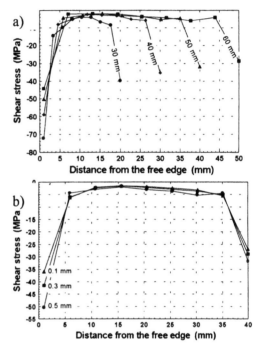

Figure 9. Adhesive sensitivity analysis. Shear stresses v a) joint length b) Adhesive thickness

is, therefore, very high for the case of carbon fibres. This coefficient reduces to lower values for the case of the aluminium leg. The maximum stresses appear in the joining region between the composite and the aluminium interface. A 3D numerical analysis of the joint region has been performed to reduce the stress to an acceptable level.

4.2 Parameters comparison

In Figure 8 are compared the most relevant design parameters for both the hybrid bar and the original steel bar. It is important to point out the reduction in weight and the significantly increase in the factor of safety due to a larger cross section.

The fatigue life for both materials is far over the customer's requirements (10^6 cycles).

4.3 Joining optimisation

The aluminium legs are bonded to the composite tube at both ends. Whilst some alternatives have been considered. Adhesive bonding is detailed here.

The transmission of loads is done through the contact composite/aluminium region. The main transmitted load is a torque of value 268.6 N·m. The joint diameter, length and adhesive thickness have been selected to give the optimum solution.

Values for the circumferencial shear stress at the adhesive mid-thickness are shown in Figure 9 for different values of the joining length and adhesive layer thickness for a given transition diameter of 22 mm.

A further increase of the length does not reduce significantly the value of the shear stresses and hence a design value of 40 mm is adopted.

5 CONCLUSIONS

A procedure for the design of an hybrid stabiliser bar for a commercial vehicle has been described. The aim of this paper has been the reduction of weight for a similar equivalent stiffness by using fibre reinforced materials.

After the conceptual analysis was performed, a bar made in medium carbon fibres impregnated with a high modulus epoxy resin has been selected. The external diameter of the bar has to be inevitably larger than in the case of steel but the cross section is hollow. The reduction in weight is over 65% which introduces an important advantage. The resulting composite bar is overdesigned with a static stress far below the ultimate strength and has a high fatigue life. The aluminium legs present a factor of safety of 2.2 which is reasonable for this kind of application.

A constant 45 degrees winding angle is calculated to be the most suitable one for this particular application due to the high torsion contribution. The displacements between both ends are the same as for the steel bar, so obtaining the same stiffness in the bar. The bending deflection along the bar is higher in the case of composite bars due to the lower bending stiffness for a winding angle of 45° degrees. The torsion stiffness can be built to be higher than that of a steel bar of 18 mm diameter, so increasing the overall bar stiffness. Therefore we would expect the composite bar to produce similar ride characteristics to that of a steel bar.

The bonded joint has been designed using finite element methods to determine the principal parameters. The maximum shear stress is reduced to 40 MPa which is lower than the ultimate shear strength for a typical high resistant epoxy adhesive.

REFERENCES

Belingardi G., Caldarale P.M. & Rosetto M. 1990. 'Design of composite material drive shafts for vehicles application. International Journal of Vehicle Design. Vol 11, 6 pp. 553.

Fenton J. Handbook of vehicle design analysis. Mechanical Engineering Publications Limited.

Goncourt L. & Sayers K.H. 1987. 'A composite automovile suspension' International Journal of Vehicle Design. Vol 8, 3 pp. 335.

Kelly A. & Rabotnov Yu.N. 1986. Handbook of composites. Vol 4: Fabrication of composites. Elservier Science Publisers B.V.

Lees W.A. 1989. Adhesives and the engineer. Pemabond Adhesive Limited.

Wienand J. 'Weight-optimised stabilisers for car axles' Extracted from "Weight-optimised spring elements for car axles". Techniseh Mitteilungen Krupp 2/95, pp. 69-78.

Woutton A.J. 1996. 'Development of a novel RP suspension system' International Journal of Vehicle Design. Vol 17, 1.

Experimental Techniques and Design In Composite Materials 4, Found (Ed.)
© 2002 Swets & Zeitlinger, Lisse, ISBN 90 5809 370 0

Strength criteria for composite material structures

A. De Iorio, D. Ianniello, R. Iannuzzi & F. Penta
Dipartimento di Progettazione e Gestione Industriale, Università di Napoli Federico II, Italy

A. Apicella & L. Di Palma
Dipartimento di Ingegneria dei Materiali e della Produzione, Università di Napoli Federico II, Italy

ABSTRACT: The development of metal matrix composites provides high specific strength materials, which are particularly necessary in the aerospace field. The complex behaviour of such materials, together with the need to provide components of high reliability, require a very deep knowledge of failure modes and material strength parameters under the load and environment conditions in which the components must operate. In the present work the authors offer a critical review of the most important design criteria which could be adopted to provide the strength and/or the stiffness qualifications for generic composite material structure, that are necessary to bear the maximum applied loads safely.

1 INTRODUCTION

The goal of every structural analysis is the evaluation of the external safety factor s_e. It is the most elevated number for which the service loads can be amplified without the structure reaches a failure condition, i.e. a condition which produce loss of utility or dangerous results.

In the following study we will consider the failure conditions produced only owing to plastic flows and/or fractures. Then, the coefficient s_e can be estimated by way of analysis if the material elastic domain or the analytical form of its boundary surface is known:

$$f\left(\sigma_{ij}, c_h\right) = 0. \quad i, j = 1,2,3; \quad h = 1,...,n; \qquad (1)$$

where σ_{ij} are the stress tensor components and c_h material parameters. For all stress states this gives:

$$f\left(\sigma_{ij}, c_h\right) < 0,$$

when only elastic strains take place, while the failure of the structure occurs when eq. (1) is verified. In such sense eq. (1) can be identified as a "failure condition" and it can be considered representative of the material strength properties.

The explicit form of eq. (1) can be drawn experimentally, executing suitable mechanical tests (phenomenological approach) and then choosing the form of the function f to give the best fit to the experimental data. Often the study of material microstructural properties and/or the analysis of experimental results lead to the formulation of a hypothesis on the failure stress states. On the other side they yield also the identification of a representative parameter of the material strength properties, because it is reasonable to assume that the failure takes place when this parameter reaches a critical value. In these cases a material failure condition is specified by a "strength criteria". This consists of a set of hypotheses which orients the choice of the explicit form of the function $f(\sigma_{ij}, c_h)$ or leads directly to the identification of this function.

2 ANISOTROPIC QUADRATIC STRENGTH CRITERIA

The most used failure conditions for anisotropic material are derivable from the criterion proposed by Von Mises (1928) to characterize the strength properties of single-crystals and having the following general formulation:

$$F_{11}\sigma_{11}^2 + F_{22}\sigma_{22}^2 + F_{33}\sigma_{33}^2 + F_{44}\sigma_{23}^2 +$$
$$+ F_{55}\sigma_{13}^2 + F_{66}\sigma_{12}^2 + 2(F_{12}\sigma_{11}\sigma_{22} + \qquad (2)$$
$$+ ... + F_{14}\sigma_{11}\sigma_{23} + ... + F_{56}\sigma_{13}\sigma_{12}) = const$$

The twenty-one independent constants F_{ij} depend upon the reference system and must be determined experimentally. However the introduction of hypotheses on symmetry of the strength properties allows to reduce their number.

2.1 Iso-strength materials

Assigned a specially orthotropic material and its reference axes $x_1 x_2 x_3$ the strength properties are symmetrical in respect of the coordinate planes, thus the quadratic form of the eq. (2) becomes:

$$F_{11}\sigma_{11}^2 + F_{22}\sigma_{22}^2 + F_{33}\sigma_{33}^2 + F_{44}\sigma_{23}^2 +$$
$$+ F_{55}\sigma_{13}^2 + F_{66}\sigma_{12}^2 + 2(F_{12}\sigma_{11}\sigma_{22} \qquad (3)$$
$$+ F_{23}\sigma_{22}\sigma_{33} + F_{13}\sigma_{11}\sigma_{33}) = const.$$

The nine parameters that appear in eq. (3) can be reduced to six if the superposition of an hydrostatic stress state does not influence the material failure. In such case, for each failure stress condition and for arbitrary values of the hydrostatic component p, the following condition must be respected by eq. (3):

$$f(\sigma_{ij}, c_h) = f(\sigma_{ij} - p\delta_{ij}, c_h) = const \qquad (4)$$

where δ_{ij} is the Kronecher symbol. So it must be:

$$p^2(F_{11} + F_{22} + F_{33} + 2F_{12} + 2F_{23} + 2F_{13}) +$$
$$+ 2 \cdot \sigma_{11}^{(f)} \cdot p \cdot (F_{11} + F_{12} + F_{13}) +$$
$$+ 2 \cdot \sigma_{22}^{(f)} \cdot p \cdot (F_{12} + F_{22} + F_{23}) +$$
$$+ 2 \cdot \sigma_{33}^{(f)} \cdot p \cdot (F_{13} + F_{23} + F_{33}) = 0$$

for arbitrary values of p and $\sigma_{ij}^{(f)}$ and this implies that:

$$\begin{cases} F_{11} + F_{22} + F_{33} + 2F_{12} + 2F_{23} + 2F_{13} = 0 \\ F_{11} + F_{12} + F_{13} = 0 \\ F_{12} + F_{22} + F_{23} = 0 \\ F_{13} + F_{23} + F_{33} = 0. \end{cases}$$

The foregoing equations can be satisfied posing:

$$F_{12} = -H; F_{13} = -G; F_{23} = -F;$$
$$F_{11} = H + G; \quad F_{22} = F + H; \quad F_{33} = G + F. \qquad (5)$$

Furthermore, identifying L, M and N as respectively the coefficients F_{44}, F_{55} and F_{66}, and using the eq. (5), then eq. (3) becomes:

$$(H + G)\sigma_{11}^2 + (F + H)\sigma_{22}^2 + (G + F)\sigma_{33}^2 +$$
$$+ 2L\sigma_{12}^2 + 2M\sigma_{23}^2 + 2N\sigma_{13}^2 +$$
$$- 2(H\sigma_{11}\sigma_{22} + F\sigma_{22}\sigma_{33} + G\sigma_{11}\sigma_{33}) = \qquad (6)$$
$$= H(\sigma_{11} - \sigma_{22})^2 + F(\sigma_{22} - \sigma_{33})^2 +$$
$$+ G(\sigma_{33} - \sigma_{11})^2 + 2L\sigma_{12}^2 + 2M\sigma_{23}^2 + 2N\sigma_{13}^2 = 1$$

that formalises the Hill criterion (1948).

The six material parameters that appear in the Hill condition can be obtained by suitable experimental tests. In fact, using respectively X, Y and Z as the normal strengths in the three principal directions x_1, x_2, x_3 and with R, S and T the shear strengths respectively in the coordinate planes x_1x_2, x_2x_3, x_1x_3, the six constants can be determined by applying eq. (6) to the six failure conditions produced to obtain the strength parameters:

$$2F = \frac{1}{Y^2} + \frac{1}{Z^2} - \frac{1}{X^2};$$
$$2G = \frac{1}{Z^2} + \frac{1}{X^2} - \frac{1}{Y^2};$$
$$2H = \frac{1}{X^2} + \frac{1}{Y^2} - \frac{1}{Z^2}; \qquad (7)$$
$$2L = \frac{1}{R^2}; 2M = \frac{1}{S^2}; 2N = \frac{1}{T^2}.$$

Hill (1950) also showed the way to specialise eq. (6) for transversally isotropic material having the x_1x_2 plane as an isotropic plane. In this case the quadratic form of eq. (6) must be invariant with respect to rotations of the coordinate system about x_3 axis. This involves:

$$F = G; N = F + 2H = G + 2H; L = M;$$

thus eq. (6) becomes:

$$H(\sigma_{11} - \sigma_{22})^2 + F(\sigma_{22} - \sigma_{33})^2 +$$
$$+ F(\sigma_{33} - \sigma_{11})^2 + +2L\sigma_{23}^2 + 2L\sigma_{13}^2 + \qquad (8)$$
$$+ 2(F + 2H)\sigma_{12}^2 = const$$

in which only four material characteristic parameters appear. If we have a biaxial stress state and the x_1x_3 plane is the stress plane, it results:

$$\sigma_{23} = \sigma_{12} = \sigma_2 = 0,$$

then eq. (8) becomes:

$$(H + F)\sigma_{11}^2 + 2F\sigma_{33}^2 +$$
$$- 2F\sigma_{11}\sigma_{33} + 2L\sigma_{13}^2 = const;$$

and using eq. (7) finally we can write:

$$\frac{\sigma_{11}^2}{X^2} + \frac{\sigma_{33}^2}{Z^2} - \frac{\sigma_{11}\sigma_{33}}{Z^2} + \frac{\sigma_{13}^2}{R} = 1. \qquad (9)$$

Eq. (9) has been used by Azzi and Tsai (1965) to describe the strength properties of a unidirectional composite.

2.2 Non Iso-strength materials

The foregoing failure conditions are valid for an iso-strength material because the quadratic forms of eqs. (2), (6) and (9) are even functions. The Von Mises condition, eq. (2), can be extended to the non iso-strength anisotropic material putting linear terms in the stress components σ_{ij} on the left side, i.e. writing eq. (2) in the form:

$$F_{ijkl}\sigma_{ij}\sigma_{kl} + F_{ij}\sigma_{ij} = const \quad i,j,k,l = 1,2,3 \qquad (10)$$

which formalizes the Tsai-Wu failure criterion (1971).

The coefficients F_{ijkl} and F_{ij} of eq. (10) are respectively the components of two symmetric tensors of the fourth and second order. In fact if K_{ijkl}, K_{ij} and K_{ijkl}, K_{ij} are the values assumed by the characteristic parameters

F_{ijkl} and F_{ij} in the two coordinate system $x_1x_2x_3$ and $x'_1x'_2x'_3$, where the stress tensor has respectively components σ_{ij} and σ'_{ij}, the failure is reached if:

$$K_{ijkl}\sigma_{ij}\sigma_{kl} + K_{ij}\sigma_{ij} =$$
$$= K'_{ijkl}\sigma'_{ij}\sigma'_{kl} + K'_{ij}\sigma'_{ij} = const \tag{11}$$

On the other hand the stress tensor components change as:

$$\sigma_{ij} = \frac{\partial x_i}{\partial x'_p}\frac{\partial x_j}{\partial x'_q}\sigma'_{pq}$$

thus, from eq. (11), the following conditions can be derived:

$$K'_{pq} = K_{ij}\frac{\partial x_p}{\partial x'_i}\frac{\partial x_q}{\partial x'_j} \tag{12}$$

$$K'_{pqhk} = K_{ijlm}\frac{\partial x_p}{\partial x'_i}\frac{\partial x_q}{\partial x'_j}\frac{\partial x_h}{\partial x'_l}\frac{\partial x_k}{\partial x'_m}, \tag{13}$$

which are the transformation laws of the second and fourth rank tensor components.

The tensorial properties of F_{ij} and F_{ijlm} make it unnecessary to deduce the symmetries of the strength properties from those of the elastic properties. In fact it is not always reasonable to suppose that those features of the microstructure that predetermine the plane of the elastic symmetry will also predetermine the plane of strength symmetry. In fact, the overall (macro) elastic properties are related to the local (micro) properties by averaging processes and are less sensitive to local variation in the microstructures than the strength properties which are highly local-fault sensitive.

Moreover, since the transformation laws of coefficients F_{ij} and F_{ijlm} are known, eqs. (12) and (13), their values can be experimentally determined in whatever coordinate system. Particularly, if X_i and X'_i are respectively the tensile and compressive strengths in the x_i direction, using eq. (10), we can write:

$$\begin{cases} F_{iiii}X_i^2 + F_{ii}X_i = 1 \\ F_{iiii}X_i^{'2} + F_{ii}X'_i = 1 \end{cases}$$

from which we have:

$$F_{iiii} = \frac{1}{X_iX'_i},$$

$$F_{ii} = \frac{1}{X_i} - \frac{1}{X'_i};$$

similarly, for the coefficients of the terms σ^2_{ij} and σ_{ij}, with $i \neq j$, it results:

$$F_{ij} = \frac{1}{X_{ij}} - \frac{1}{X'_{ij}} \tag{14}$$

$$F_{ijij} = \frac{1}{X_{ij}X'_{ij}} \tag{15}$$

where X_{ij} and X'_{ij} are respectively the positive and negative shear strengths in the direction x_i and x_j.

Finally the coefficients of the rectangular terms F_{ijlm} with $(i,j)\neq(l,m)$ can be determined by biaxial tests in which the material reaches the failure owing to a stress state where only the components σ_{ij} and σ_{lm} are not zero. In fact if $\sigma^{(f)}_{ij}$ and $\sigma^{(f)}_{lm}$ are respectively the failure values of σ_{ij} and σ_{lm}, using eq. (10) we have:

$$g(\sigma^{(f)}_{ij},\sigma^{(f)}_{lm},F_{ijlm}) =$$
$$= F_{ijij}\sigma^{(f)2}_{ij} + F_{lmlm}\sigma^{(f)2}_{lm} + \tag{16}$$
$$+ 2F_{ijlm}\sigma^{(f)}_{ij}\sigma^{(f)}_{lm} + F_{ij}\sigma^{(f)}_{ij} + F_{lm}\sigma^{(f)}_{lm} = 1$$

from which we can obtain the following expression of F_{ijlm}:

$$F_{ijlm} = -\frac{1 - F_{ijij}\sigma^{(f)2}_{ij} + F_{lmlm}\sigma^{(f)2}_{lm} + F_{ij}\sigma^{(f)}_{ij} + F_{lm}\sigma^{(f)}_{lm}}{2\sigma^{(f)}_{ij}\sigma^{(f)}_{lm}}.$$

If the test is carried out maintaining constant the ratio $B = \sigma_{lm}/\sigma_{ij}$ until the material fails, we can write:

$$F_{ijlm} = -\frac{1 - F_{ijij}\sigma^{(f)2}_{ij} + F_{lmlm}B\sigma^{(f)2}_{ij} + F_{ij}\sigma^{(f)}_{ij} + F_{lm}B\sigma^{(f)}_{ij}}{2B\sigma^{(f)2}_{ij}}.$$
$$\tag{17}$$

On the other hand, since the failure envelope, eq. (10), must be a real closed surface, it needs that the coefficients F_{ijij}, F_{lmlm} and F_{ijlm} respect the stability condition:

$$F_{ijij}F_{lmlm} - F^2_{ijlm} \leq 0,$$

which corresponds to the following limitations for the value of the coefficient F_{ijlm}:

$$-\sqrt{F_{ijij}F_{lmlm}} \leq F_{ijlm} \leq +\sqrt{F_{ijij}F_{lmlm}}. \tag{18}$$

The scatter of experimental data sometimes can leads to F_{ijlm} estimations which do not respect the stability conditions, eq. (18). If $\delta\sigma^{(f)}_{ij}$ is the scatter of the failure value $\sigma^{(f)}_{ij}$, obtained for a fixed value of the ratio B, the consequent absolute scatter of F_{ijlm} will be valuable with the following expression:

$$\delta F_{ijlm} = \frac{\partial F_{ijlm}}{\partial\sigma^{(f)}_{ij}}\delta\sigma^{(f)}_{ij} =$$
$$-\left(\frac{F_{ijij}}{B} + F_{lmlm}B + \frac{F_{ij}}{2B} + \frac{F_{lm}}{2\sigma^{(f)}_{ij}} + 2F_{ijlm}\right)\frac{\delta\sigma^{(f)}_{ij}}{\sigma^{(f)}_{ij}} \tag{19}$$

or, in terms of relative scatter, by means of the expression:

$$\frac{\delta F_{ijlm}}{F_{ijlm}} = -\psi(\sigma^{(f)}_{ij},B)\frac{\delta\sigma^{(f)}_{ij}}{\sigma^{(f)}_{ij}}$$

in which the magnification factor ψ is:

$$\psi(\sigma_{ij}^{(f)}, B) = -\frac{1}{F_{ijlm}} \cdot$$

$$\cdot \left(\frac{F_{ijij}}{B} + F_{lmlm}B + \frac{F_{ij}}{2B} + \frac{F_{lm}}{2\sigma_{ij}^{(f)}} + 2F_{ijlm} \right)$$

Thus to respect the condition (18) the failure values of the stress components must be such that:

$$\delta F_{ijlm}\left(\sigma_{ij}^{(f)}, B\right) << \sqrt{F_{ijij}F_{lmlm}} \ .$$

Wu (1972) proposes to evaluate F_{ijlm} by means of the values of $\sigma_{ij}^{(f)}$ and B such where the function ψ reaches a minimum, that is:

$$\frac{d\psi}{dB} = \frac{\partial \psi}{\partial B} + \frac{\partial \psi}{\partial \sigma_{ij}^{(f)}} \frac{\partial \sigma_{ij}^{(f)}}{\partial B} = 0, \qquad (20)$$

where the derivative $\dfrac{\partial \sigma_{ij}^{(f)}}{\partial B}$ can be obtained from the failure conditions eq. (16). Thus, the condition (20) becomes:

$$\frac{2\sigma_{ij}^{(f)}\left(F_{lmlm}B + F_{ijlm}\right) + F_{lm}}{2\sigma_{ij}^{(f)}(F_{lmlm}B^2 - F_{ijij}) - F_{ij}} =$$
$$= \frac{2\sigma_{ij}^{(f)}(F_{lmlm}B^2 + 2F_{ijlm}B + F_{ijij}) + F_{lm}B + F_{ij}}{-B(BF_{lm} + F_{ij})} \qquad (21)$$

that can be simultaneously solved with eq. (16) (the latter denotes the relation between $\sigma^{(f)}_{ij}$ and B). The coefficient F_{ijlm} of eqs. (16) and (21) is unknown, therefore those equations can be simultaneously solved by means of an iterative procedure, when it is assigned an initial approximate value of F_{ijlm} respecting the condition (18). For that reason it needs to trace out the δF_{ijlm}-F_{ijlm} curves, with $F_{ijlm} \in \left[- \sqrt{F_{ijij}F_{lmlm}} \ ; \sqrt{F_{ijij}F_{lmlm}} \ \right]$, at different values of B, by which it's possible individuate a value B^* of B such that the following condition is verified:

$$\delta F_{ijlm} << \sqrt{F_{ijij}F_{lmlm}}$$

$$\forall \ F_{ijlm} \in \left[- \sqrt{F_{ijij}F_{lmlm}} \ , \sqrt{F_{ijij}F_{lmlm}} \ \right]$$

The initial value of F_{ijlm} can be calculated by eq. (17), using the failure value $\sigma^{(f)}_{ij}$ obtained from a test carried out with $B=B^*$. So the eqs. (16) and (21) can be solved yielding a second value B^{**} of B. The results of a test carried out with $B=B^{**}$ allows us to calculate a second value of F_{ijlm}. The foregoing operations must be iterated until the values of F_{ijlm}, B and $\sigma^{(f)}_{ij}$ are such that the eq. (21) is verified with adequate approximation.

The number of material constants is reduced if strength property symmetries can be detected. If the material is orthotropic, besides the coefficients of the rectangular terms containing the product between a symmetric and an asymmetric stress component with respect to the $x_i x_j$ plane, also the coefficients of the linear terms in asymmetric stress component must be zero. Therefore, the Hill condition can be used to represent the strength properties of a non iso-strength orthotropic material if it is written in the form:

$$H(\sigma_{11} - \sigma_{22})^2 + F(\sigma_{22} - \sigma_{33})^2 +$$
$$+ G(\sigma_{33} + -\sigma_{11})^2 + + 2L\sigma_{12}^2 + 2M\sigma_{23}^2 + \qquad (22)$$
$$+ 2N\sigma_{13}^2 + U\sigma_{11} + V\sigma_{22} + W\sigma_{33} = const$$

where the parameters U, V and W, to respect the condition (4), must be such that U+V+W=0. Eq. (24) is the Hoffmann failure criterion (1967).

3 ENERGY CRITERIA

If the stress tensor $\underline{\sigma}$, having components σ_{ij}, is arbitrarily decomposed into two tensors $\underline{\sigma}'$ and $\underline{\sigma}''$, whose components respectively are σ_{ij} and σ_{ij}, it results:

$$\sigma_{ij} = \sigma'_{ij} + \sigma''_{ij} \qquad (23)$$

and the elastic strain energy can be expressed in the following form:

$$U = \frac{1}{2}\sigma_{ij}\varepsilon_{ij} = \frac{1}{2}\left(\sigma'_{ij} + \sigma''_{ij}\right)\left(\varepsilon'_{ij} + \varepsilon''_{ij}\right) =$$
$$= \frac{1}{2}\sigma'_{ij}\varepsilon'_{ij} + \frac{1}{2}\sigma''_{ij}\varepsilon''_{ij} + \frac{1}{2}\left(\sigma'_{ij}\varepsilon''_{ij} + \sigma''_{ij}\varepsilon'_{ij}\right)$$

where ε'_{ij} and ε''_{ij} are respectively the components of the strain states $\underline{\varepsilon}'$ and $\underline{\varepsilon}''$ associated with $\underline{\sigma}'$ and $\underline{\sigma}''$. Since the elastic compliance tensor $[S_{ijkl}]$ is symmetric, we have:

$$\sigma'_{ij}\varepsilon''_{ij} = \sigma'_{ij}S_{ijlm}\sigma''_{lm} = \sigma''_{lm}S_{lmij}\sigma'_{ij} = \sigma''_{lm}\varepsilon'_{lm}$$

therefore the elastic strain energy can be expressed as:

$$U = \frac{1}{2} \cdot \sigma'_{ij} \cdot \varepsilon'_{ij} + \frac{1}{2} \cdot \sigma''_{ij} \cdot \varepsilon''_{ij} + \sigma'_{ij}\varepsilon''_{ij}$$

Choosing the tensors $\underline{\sigma}'$ and $\underline{\sigma}''$ respectively equal to the hydrostatic part $\underline{\sigma}^{(v)}$ and the deviatoric part $\underline{\sigma}^{(d)}$ of $\underline{\sigma}$, we have:

$$\sigma'_{ij} = \sigma_{ij}^{(v)} = p\delta_{ij}$$
$$\sigma''_{ij} = \sigma_{ij}^{(d)} = \sigma_{ij} - \sigma'_{ij}$$

where $\sigma_{ij}^{(v)}$ are the $\underline{\sigma}^{(v)}$ components and $\sigma_{ij}^{(d)}$ the $\underline{\sigma}^{(d)}$ components. When the elastic properties are isotropic, the $\underline{\varepsilon}'$ components are $\varepsilon_{ij} = \varepsilon_v\delta_{ij}$, with $\varepsilon_v = (\varepsilon_{11} + \varepsilon_{22} + \varepsilon_{33})/3$, so this tensor is hydrostatic and yield a volume change equal to that yielded by the stress state $\underline{\sigma}$. Furthermore the tensor $\underline{\varepsilon}''$ is deviatoric because it results $\varepsilon''_{11} + \varepsilon''_{22} + \varepsilon''_{33} = 0$ and it corresponds to a pure distortion. Since it results:

$$\sigma''_{ij}\varepsilon'_{ij} = \sigma_{ij}^{(d)}\varepsilon_{ij}^{(v)} = \left(\sigma_{11}^{(d)} + \sigma_{22}^{(d)} + \sigma_{33}^{(d)}\right)\varepsilon_v = 0,$$

the strain energy is:

$$U = \frac{1}{2}\sigma_{ij}^{(v)}\varepsilon_{ij}^{(v)} + \frac{1}{2}\sigma_{ij}^{(d)}\varepsilon_{ij}^{(d)}.$$

Thus, U is the sum of volume change energy $U_v = (1/2)(\sigma_{ij}^{(v)}\varepsilon_{ij}^{(v)})$ and of elastic distortion energy $U_d = (1/2)(\sigma_{ij}^{(d)}\varepsilon_{ij}^{(d)})$. When the material is specially orthotropic, posing $\sigma'_{ij} = \sigma_{ij}^{(v)} = p\delta_{ij}$, the strain components ε'_{ij} identify a volume change different from the effective one. Therefore the stress state $\sigma''_{ij}=\sigma_{ij}-\sigma_{ij}$ is deviatoric but does not yield only distortion. If $\underline{\sigma}'$ is a particular stress state producing only the effective hydrostatic strain state, it means that $\underline{\sigma}'$ is not hydrostatic and $\underline{\sigma}''$ is not deviatoric while the strain state $\underline{\varepsilon}''$ is a pure distortion. In any case the elastic strain energy cannot be evaluated by adding only the terms $(1/2)(\sigma'_{ij}\varepsilon'_{ij})$ and $(1/2)(\sigma''_{ij}\varepsilon''_{ij})$.

The decomposition of the stress tensor into hydrostatic and deviatoric parts is useful to study the strength properties of materials reaching failure owing to phenomena activated by shear stress, such as metallic materials, in which macroscopic plastic flow starts when the shear stress components, acting on the slipping planes, reach a critical value. Since the hydrostatic part does not produce shear stress the failure condition can be expressed in term of only the deviatoric stress components. On the other hand if it is reasonable to assume that failure take place when the deviatoric stress components work, W_d, reaches a characteristic value, in the isotropic case it results in $W_d = U_d$ and the elastic distortion energy can be chosen as a parameter representative of the material strength properties.

Griffith and Baldwin (1962) assume that the material failure takes place when an opportune fraction of the elastic strain energy, improperly termed distortion strain energy, reaches a characteristic value. The work carried out by the stress components during the volume change is:

$$U'_v = \frac{\sigma_{11} + \sigma_{22} + \sigma_{33}}{2}\varepsilon_v.$$

The strain energy fraction U_l producing the material failure is obtained by subtracting U'_v from the total strain energy:

$$U_1 = U - U'_v;$$

therefore the failure condition proposed by Griffith and Baldwin is:

$$U_1 = \sigma_{11}^2\left(\frac{S_{11}}{3} - \frac{S_{12}}{6} - \frac{S_{13}}{6}\right) + \sigma_{22}^2\left(\frac{S_{22}}{3} - \frac{S_{12}}{6} - \frac{S_{23}}{6}\right) +$$
$$\sigma_{33}^2\left(\frac{S_{33}}{3} - \frac{S_{13}}{6} - \frac{S_{23}}{6}\right) + +\sigma_{11}\sigma_{22}\left(\frac{2S_{12}}{3} - \frac{S_{11}}{6} - \frac{S_{22}}{6} +\right.$$
$$\left. - \frac{S_{13}}{6} - \frac{S_{23}}{6}\right) + \sigma_{22}\sigma_{33}\left(\frac{2S_{23}}{3} - \frac{S_{12}}{6} - \frac{S_{22}}{6} - \frac{S_{13}}{6} +\right.$$
$$\left. - \frac{S_{33}}{6}\right) + \sigma_{11}\sigma_{33}\left(\frac{2S_{13}}{3} - \frac{S_{11}}{6} - \frac{S_{12}}{6} - \frac{S_{23}}{6} - \frac{S_{33}}{6}\right) +$$
$$+ \sigma_{12}^2 S_{66} + \sigma_{23}^2 S_{44} + \sigma_{13}^2 S_{55} = U_1^\sigma$$

where $U_1^{(cr)}$ is the critic value of U_1.

In order to evaluate the elastic energy fraction which yield failure, Formann (1972), decompose the stress tensor into the parts $\underline{\sigma}^*$ and $\underline{\sigma}^{**}$ where the first is the stress tensor associated with hydrostatic part of strain tensor. In the principal coordinate system of an orthotropic material it results in $\sigma_{ij} = G_{ij}\varepsilon_{ij}$, with $i \ne j$, G_{ij} being the shear elastic modulus, then the normal components of $\underline{\sigma}^*$ can be calculated, solving the following algebraic system:

$$\begin{cases} \dfrac{\overset{*}{\sigma}_{11}}{E_{11}} - v_{12}\dfrac{\overset{*}{\sigma}_{22}}{E_{22}} - v_{13}\dfrac{\overset{*}{\sigma}_{33}}{E_{33}} = \varepsilon_{11}^v \\[2mm] -v_{12}\dfrac{\overset{*}{\sigma}_{11}}{E_{11}} - \dfrac{\overset{*}{\sigma}_{22}}{E_{22}} - v_{23}\dfrac{\overset{*}{\sigma}_{33}}{E_{33}} = \varepsilon_{22}^v \\[2mm] -v_{13}\dfrac{\overset{*}{\sigma}_{11}}{E_{11}} - v_{23}\dfrac{\overset{*}{\sigma}_{22}}{E_{22}} - \dfrac{\overset{*}{\sigma}_{33}}{E_{33}} = \varepsilon_{33}^v \end{cases}$$

where E_{ii} are the Young's modulus in the principal directions and v_{ij} the Poisson's ratio.

So the strain energy can be written in terms of the components of $\underline{\sigma}^*$, $\underline{\sigma}^{**}$ and $\varepsilon^{(v)}$, $\varepsilon^{(d)}$ as:

$$U = \frac{1}{2}\overset{*}{\sigma}_{ij}\varepsilon_{ij}^{(v)} + \frac{1}{2}\overset{**}{\sigma}_{ij}\varepsilon_{ij}^{(d)} + \overset{*}{\sigma}_{ij}\varepsilon_{ij}^{(d)}.$$

Formann assumes that the material reaches failure when the strain energy fraction:

$$U_2 = \frac{1}{2}\overset{**}{\sigma}_{ij}\varepsilon_{ij}^{(d)} + \overset{*}{\sigma}_{ij}\varepsilon_{ij}^{(d)},$$

that is the distortion energy, reaches the critical value $U_2^{(cr)}$.

Pagano carries out the decomposition of the stress tensor imposing some additional conditions. If σ_{ij}, σ_{ij} and ε_{ij}, ε_{ij} are the components of the parts in which the stress and the strain tensors have to be decomposed, they are defined by means of the eq. (23) and by the following relationships:

$$S_{ijlm}\sigma'_{ij}\sigma''_{lm} = 0$$
$$\varepsilon''_{11} + \varepsilon''_{22} + \varepsilon''_{33} = 0$$
$$\sigma''_{11} + \sigma''_{22} + \sigma''_{33} = 0.$$

The first of the foregoing conditions assures that the mutual work is zero, while the remaining conditions assure that $\underline{\sigma}''$ and $\underline{\varepsilon}''$ are deviatoric tensors. In such case the elastic strain energy U is equal to $(1/2)(\sigma'_{ij}\varepsilon'_{ij}) + (1/2)(\sigma''_{ij}\varepsilon''_{ij})$, and the quantity $(1/2)(\sigma''_{ij}\varepsilon''_{ij})$ is the distortion energy U_d. If U_d is responsible for the material failure, the elastic domain boundary has the expression:

$$U_d = \frac{1}{2}\sigma''_{ij}\varepsilon''_{ij} = U_d^{cr}$$

where U_d^{cr} is the critic value of U_d.

4 LAMINATE STRENGTH ANALYSIS

The method usually employed to verify the strength of composite laminates is the "ply-to-ply" analysis. If the characteristics of internal reactions N_{ij}, T_{ij} and M_{ij} or the generalized strains $\varepsilon_{ij}^{(0)}$ and k_{ij} are known the stress component $\sigma_{ij}^{(k)}$ acting on the lamina k of the plate element can be evaluated by means of the *lamination theory*. Thus it is simple to check the strength of lamina k if its failure envelope is known.

When the laminate is symmetric and is subjected to a membrane loading condition, the strength analysis can be easily carried out expressing the failure condition of the generic lamina directly in terms of the characteristics of internal reactions of the laminate (Chou et al., 1973). In this case the stress state of the lamina k is uniform through the thickness and the generic component $\sigma_{ij}^{(k)}$ can be expressed by the formula:

$$\sigma_{ij}^{(k)} = \frac{1}{h} U_{ijpq}^{(k)} N_{pq} \qquad (24)$$

where N_{pq} are the membrane characteristics and h is the laminate thickness. The $U_{ijlm}^{(k)}$ coefficients can be obtained by inverting the laminate and the lamina elastic constitutive equations:

$$\varepsilon_{ij}^{(0)} = A_{ijlm} N_{lm}$$
$$\varepsilon_{ij}^{(k)} = S_{ijlm}^{(k)} \sigma_{lm}^{(k)}$$

where A_{ijlm} and $S_{ijlm}^{(k)}$ are respectively the elastic compliance of the plate and of the lamina. If the strength properties of lamina k are expressed by the Tsai-Wu failure condition:

$$F_{ijlm}^{(k)} \sigma_{ij}^{(k)} \sigma_{lm}^{(k)} + F_{ij}^{(k)} \sigma_{ij}^{(k)} = 1 \,,$$

substituting in this equation the eq. (24), in which we pose $\bar{\sigma}_{pq} = N_{pq}/h$, we obtain:

$$\bar{F}_{pqts}^{(k)} \bar{\sigma}_{pq} \bar{\sigma}_{ts} + \bar{F}_{pq}^{(k)} \bar{\sigma}_{pq} = 1$$

where $\bar{F}_{pqts}^{(k)} = F_{ijlm}^{(k)} U_{ijpq}^{(k)} U_{lmts}^{(k)}$ and $\bar{F}_{pq}^{(k)} = F_{ijlm}^{(k)} U_{ijpq}^{(k)}$. The previous equation specifies the laminate loading conditions which yield failure in the lamina k.

It provides a different approach to characterise the strength of symmetric laminates subjected to membrane actions. This is the direct characterisation in which the laminate is regarded as a homogeneous material whose strength properties can be expressed by a quadratic form in the components $\bar{\sigma}_{ij}$ or using the tensorial polynomial formulation proposed by Wu and Scheublein (1974):

$$f(\bar{\sigma}_{ij}) = F_{ij} \bar{\sigma}_{ij} + F_{ijhk} \bar{\sigma}_{ij} \bar{\sigma}_{hk} + F_{ijhklm} \bar{\sigma}_{ij} \bar{\sigma}_{hk} \bar{\sigma}_{lm} = 1$$
with $i, j, l, m, h, k = x, y$
$$(25)$$

where F_{ij}, F_{ijhk} and F_{ijhklm} are respectively the components of second, fourth and sixth order tensors. This approach does not require the knowledge of the strength properties of every lamina and the determination of the stress distribution through the laminate thickness. The components of the three strength tensors F_{ij}, F_{ijhk} and F_{ijklm} are estimated experimentally by executing mechanical tests on the whole laminate. For components F_{ij} and F_{ijhk} the consideration exposed in the section 2.2 are still valid. The sixth order strength tensor has twenty-seven independent coefficients which must be determined on the basis of experimental results. If the failure is loading path independent, the function $f(\bar{\sigma}_{ij})$ of eq. (25) must be a potential function of C3 class, i.e. it must be:

$$F_{ijhklm} = \frac{\partial^3 f}{\partial \bar{\sigma}_{ij} \partial \bar{\sigma}_{hk} \partial \bar{\sigma}_{lm}} = \frac{\partial^3 f}{\partial \bar{\sigma}_{ij} \partial \bar{\sigma}_{lm} \partial \bar{\sigma}_{hk}} = F_{ijlmhk}$$

$$F_{ijhklm} = \frac{\partial^3 f}{\partial \bar{\sigma}_{ij} \partial \bar{\sigma}_{hk} \partial \bar{\sigma}_{lm}} = \frac{\partial^3 f}{\partial \bar{\sigma}_{lm} \partial \bar{\sigma}_{hk} \partial \bar{\sigma}_{ij}} = F_{lmhkij}$$

thus we have the following equations:

$$F_{ijhklm} = F_{ijlmhk} = F_{hkijlm} = F_{hklmij} = F_{lmijhk} = F_{lmhkij}$$

by which the total number of independent coefficients is reduced to ten. Moreover, if the laminate is specially orthotropic with respect to the x and y axes the third order terms which are odd functions of the shear stress component must be zero, that is:

$$F_{xxxxxy} = F_{xxyyxy} = F_{yyyyxy} = F_{xyxyxy} = 0$$

Finally if the coefficients F_{ij} and F_{ijij} are calculated by means of formulas (14) and (15), it must be $F_{ijijij}=0$. So to characterize the strength properties of and orthotropic laminate subjected to membrane actions it need evaluate the following parameters:

$$F_{xx}, F_{yy}, F_{xxx}, F_{yyy}, F_{xyxy}, F_{xxyy},$$
$$F_{xxyyyy}, F_{xxxxyy}, F_{yyxyxy}, F_{xxxyxy}$$

The value of F_{xxyy} is determined on the basis of the results of mechanical tests in which the material reaches failure owing to a stress state having only the component $\bar{\sigma}_{xx}$ and $\bar{\sigma}_{yy}$ different to zero. If $\bar{\sigma}_{xx}^{(f)}$ and $\bar{\sigma}_{yy}^{(f)}$ are their failure values, the eq. (25) becomes:

$$\bar{\sigma}_{xx}^{(f)3} (3F_{xxxxyy}B + 3F_{xxyyyy}B^2) + \bar{\sigma}_{xx}^{(f)2} (F_{xxxx} +$$
$$+ F_{yyyy}B^2 + +2F_{xxyy}B) + \bar{\sigma}_{xx}^{(f)} (F_{xx} + F_{yy}B) = 1 \qquad (26)$$

where $B = \dfrac{\sigma_{yy}^{(f)}}{\sigma_{xx}^{(f)}}$; thus we have the following expression of F_{xxyy}:

$$F_{xxyy} = \frac{1 - \bar{\sigma}_{xx}^{(f)3} (3F_{xxxxyy}B + 3F_{xxyyyy}B^2)}{2B\bar{\sigma}_{xx}^{(f)2}} +$$

$$- \frac{\bar{\sigma}_{xx}^{(f)2} (F_{xxxx} + F_{yyyy}B^2) - \bar{\sigma}_{xx}^{(f)} (F_{xx} + F_{yy}B)}{2B\bar{\sigma}_{xx}^{(f)2}}.$$

Also in this case the value of ratio B must be chosen in way to minimize the scatter function δF_{xxyy}:

$$\delta F_{xxyy} = \frac{\partial F_{xxyy}}{\partial \overline{\sigma}_{xx}^{(f)}} \cdot \delta \overline{\sigma}_{xx}^{(f)} = \psi_{xxyy} \cdot \frac{\delta \overline{\sigma}_{xx}^{(f)}}{\overline{\sigma}_{xx}^{(f)}}$$

where the magnification factor ψ_{xxyy} is:

$$\psi_{xxyy} = -\frac{2 + \sigma_{xx}^{(f)3}(3F_{xxxyy}B + 3F_{xxyyyy}B^2)}{2B\sigma_{xx}^{(f)2}} + $$

$$-\frac{\sigma_{xx}^{(f)}(F_{xx} + F_{yy}B)}{2B\sigma_{xx}^{(f)2}}$$

Thus, the optimum ratio for B is valuable for solving the algebraic system formed by eq. (26) and the stationary condition of the function ψ_{xxyy}:

$$\left.\frac{d\psi_{xxyy}}{dB}\right|_{F_{xxyy}} = \frac{\partial \psi_{xxyy}}{\partial B} + \frac{\partial \psi_{xxyy}}{\partial \overline{\sigma}_{xx}^{(f)}}\frac{d\overline{\sigma}_{xx}^{(f)}}{dB} = 0 \quad (27)$$

In these equations besides the coefficients F_{xx}, F_{yy}, F_{xxxx} and F_{yyyy}, whose values are calculated on the ground of compressive and tensile strengths in the principal direction by means of eqs. (14) and (15), also the unknown coefficients F_{xxyy}, F_{xxxyy} and F_{xxyyyy} appear. Even the value of F_{xxyyyy} and F_{xxyyyy} must be evaluated by an analogous procedure to that exposed for F_{xxyy}. It consists of substituting in their expressions, deduced from eq. (26), the failure value of the stress component $\sigma_{xx}^{(f)}$ obtained by mechanical tests in which the ratio of B minimises the scatter functions δF_{xxxyy} and δF_{xxyyyy} and consequently the magnification factors:

$$\psi_{xxxyy} = \frac{\delta F_{xxxyy}}{\frac{\delta \overline{\sigma}_{xx}^{(f)}}{\overline{\sigma}_{xx}^{(f)}}} = \frac{\overline{\sigma}_{xx}^{(f)2}(F_{xxxx} + F_{yyyy}B^2 + 2F_{xxyy}B)}{3B\overline{\sigma}_{xx}^{(f)3}} +$$

$$+\frac{2\overline{\sigma}_{xx}^{(f)}(F_{xx} + F_{yy}B) - 3}{3B\overline{\sigma}_{xx}^{(f)3}} = 0$$

$$\psi_{xxyyyy} = \frac{\delta F_{xxyyyy}}{\frac{\delta \overline{\sigma}_{xx}^{(f)}}{\overline{\sigma}_{xx}^{(f)}}} = \frac{\overline{\sigma}_{xx}^{(f)2}(F_{xxxx} + F_{yyyy}B^2 + 2F_{xxyy}B)}{3B^2\overline{\sigma}_{xx}^{(f)3}} +$$

$$+\frac{2\overline{\sigma}_{xx}^{(f)}(F_{xx} + F_{yy}B) - 3}{3B^2\overline{\sigma}_{xx}^{(f)3}} = 0$$

Therefore it is convenient to solve simultaneously three algebraic systems, each of them being made by eq. (26) and one of the stationary conditions of the amplification factors ψ_{xxyy}, ψ_{xxxyy} and ψ_{xxyyyy}. There can be solved by means of an iterative procedure starting from approximate initial values of F_{xxyy}, F_{xxxyy} and F_{xxyyyy}.

So three values of the ratio B with which we carry out the tests for a better estimate of the unknown coefficients are obtained. The iteration can go on until the stationary condition of magnification functions ψ_{xxyy}, ψ_{xxxyy} and ψ_{xxyyyy} are satisfied with adequate approximation. The same procedure must be used for F_{xxxyy} and F_{yyxyy}.

5 CONCLUSION

The present work represents an attempt to examine closely the most used failure criteria to characterise the strength properties of anisotropic composite materials. The authors carried out a detailed analysis in order to underline the basic hypotheses of each formulation. They also showed which features complicate the strength criteria formulation only on the basis of energy considerations and the procedures to characterise the strength properties of laminates starting from the lamina properties.

The authors cannot give any general judgement about the validity of the several criteria discussed. Moreover, no direction can be suggested about the choice of the most suitable one. In fact, the choice depends on the criterion most suitable for fitting the experimental data. Nevertheless the analysis showed the operational limits of the Tsai-Wu and Wu-Scheublein failure conditions which are derived under the less restrictive hypotheses. The material parameters which appear in these criteria must be determined by complex and expensive iterative procedures, whose convergence properties are at present unknown. The remaining criteria, have been derived under more restrictive hypotheses about the material strength properties and the failure stress states, but they can be applied using the results of the usual mechanical tests.

REFERENCES

Azzi D., and Tsai S.W., (1965). Anisotropic Strength of Composites, Exp. Mech, 5, p. 283.

Chou P.C., Mc Namee B.M., and Chou, (1973). D. K., The Yield Criterion of Laminated Media, J. Comp. Mat., 7, p. 22.

Forman G.W., (1972). A Distortional Energy Failure Theory for Orthotropic Materials, J. Eng. Ind., 94, p. 1073.

Griffith J.E., and Baldwin W.M., (1962). Failure Theories for Generally Orthotropic Materials, Proceedings of the 1st Southern Conference on Theoretical and Applied Mechanics, p. 410.

Hill R., (1948). Proc. Roy. Soc. A, 193, p. 281.

Hill R., (1950). The Mathematical Theory of Plasticity, Clarendon Press, Oxford.

Hoffman O., (1967). The Brittle Strength of Orthotropic Materials, J. Comp. Mat., I. p. 200.

Mises R. Yon, (1928), Z. Angew. Math. Mech., 8, pp. 161-185.

Pagano NJ., (1975). Distortional Energy of Composite Materials, J. Comp. Mat., 9, p. 67.

Tsai S.W., and Wu E.M., (1971). A General Theory of Strength for Anisotropic Materials, J. Comp. Mat., 5, p. 58.

Wu E.M., (1972). Optimal Experimental Measurement of Anisotropic Failure Tensors, J. Comp. Mat., 6, p. 472.

Wu E.M., and Scheublein, J.K., (1974). *Laminate Strength – A Direct Characterization Procedure*, Composite Materials; Testing and Design, ASTM STP 546, p. 188.

Experimental Techniques and Design In Composite Materials 4, Found (Ed.)
© *2002 Swets & Zeitlinger, Lisse, ISBN 90 5809 370 0*

On the static axial collapse of square composite rail vehicle hollow bodyshells

A.G. Mamalis, D.E. Manolakos, M.B. Ioannidis & P.K. Kostazos
Manufacturing Technology Division, National Technical University of Athens, Greece
M. Robinson
Advanced Railway Research Centre, University of Sheffield, U.K.

ABSTRACT: Some preliminary experimental results pertaining to the static axial loading of hybrid square rail vehicle tubular components, made of foam-cored composite sandwich panels with integral energy absorbing inserts, are reported. The two structural configurations tested are the "tubular" and the "corrugated" core systems. Failure modes at macro- and microscale and the energy absorbing capability of the collapsed structural components are presented. The crashworthy behaviour of these small-scale bodyshells seems to be greatly affected by the structural design and the material properties of the composite sandwich components.

1 INTRODUCTION

In vehicle design, during the last decade, attention has been directed towards light weighting, life cycle costing and crashworthiness. Crash energy management has been focused on composite structures, designed to collapse in a controlled manner, which dissipates safely the kinetic energy and limits the accelerations transmitted to the occupants, aiming at the possible partial replacement of "traditional" engineering materials, such as steel and aluminium. The main advantages of composite materials over more conventional materials are the high specific strengths and specific stiffnesses which can be achieved. Composites can be designed to provide collision energy absorption capabilities which are superior to those of metals when compared on a weight-for-weight basis [1, 2].

Fibre-reinforced plastics do not exhibit the ductile failure modes associated with metals, instead, the brittle nature of most fibres and resins tends to generate brittle microfailures, such as matrix cracking, delamination, fibre breakage, etc. which constitute the main failure modes of the collapsed structures. Provided that the crushing mechanisms can be controlled, so that the composite material fails in a stable progressive manner, high levels of energy can be absorbed. Note, however, that crashworthiness cannot be at the expense of other safety issues, e.g. fire safety, therefore, selection of composite materials must take this into account. Moreover, with composites, the designer can vary the type of fibre, matrix and fibre orientation to produce composites with improved material properties [1, 2].

Extensive theoretical and experimental research work has been performed on axial loading of thin- (or thick-) walled composite structures. The effect of specimen geometry on the energy absorption capability was investigated by varying the cross-sectional dimensions, wall thickness and length of the shell. The effect of the type of composite material, laminate design, loading method and strain-rate on the crashworthy behaviour as well as enviromental effects related to crash characteristics of composites, have been also studied. An extensive literature survey on the topic may be found in Refs. [1] and [2], but see also Ref. [3].

Apart from the fibre reinforced plastics, sandwich composite material structures are of great importance. As in the case of fibre reinforced plastics, sandwich structures exhibit considerable energy absorption potential. Energy can be effectively absorbed though both local core crushing and global deflection of the panel as a whole. The energy absorption capability is largely controlled by the nature of the core, the face plates and the adhesive system used, as well as by the overall geometry of the panels and the supporting techniques employed [4].

In the present paper are reported some preliminary experimental results pertaining to the crashworthy behaviour of square, relatively thick, rail vehicle tubular components when subjected to static axial collapse. They consist of foam-cored sandwich panels with integral energy absorbing inserts; two different hybrid systems, designated as "tubular" and "corrugated" core systems, were fabricated. The failure modes at macro- and microscopic scale, as well as the energy absorbing characteristics of the collapsed components are presented and discussed.

2 EXPERIMENTAL

Hollow square structures, fabricated from composite sandwich panels tied with internal fibre reinforced plastic cylindrical tubes or corrugations, have been tested under quasi-static axial compression. Two different cross-sections were fabricated, designated as material A and B, respectively; their structural details are shown in Fig. 1.

Material A is a sandwich composite material consisting of two fibreglass face plates constructed from two layers of 1168 g/m^2 [0/45/90/-45] non-crimp quadriaxial mat (placed back-to-back to produce a symmetric laminate) impregnated in phenolic resin and a rigid polyurethane foam core material of 130 kg/m^3 density. The thickness of the face plates is 2 mm, whilst a 40 g/m^2 surface veil is placed on the outer surfaces. Both the internal and external face plates are tied to each other by cylindrical tubes of 25 mm length, formed from 25 mm diameter [+/-45] glass fibres, braided in the same resin as the face plates. Additionally, a layer of 450 g/m^2 continuous fibre mat/chopped strand mat is placed on either side of the foam core, before the quadraxial facings. The ends of the braided inserts then pass through this random fibre layer, improving, in this manner, the bond strength between the ends of the braided inserts and the face plate laminates. A pair of cylindrical inserts is centered along the overall width of each side wall of the hollow square specimen with a spacing of 70 mm. Pairs of inserts are spaced at 70 mm down the entire height of each side wall of the specimen, but the inserts of each pair are alternatively placed with a spacing of 35 mm to each other along the axis of the specimen. The height of the cell, with the two parallel top and bottom ends ground square, was 450 mm.

Material B has exactly the same geometry, face plates and foam-core as the Material A, but instead of the layer of 450 g/m^2 and the cylindrical fibre reinforced plastic inserts of Material A, Material B consists of a corrugated fibreglass laminate of laying-up similar to that of the face plates, bonded to them at distinct positions and crossing internally the foam-core all over the tube circumference; see Fig. 1.

The static collapse was carried out between the parallel steel platens of an SMG 1000 kN hydraulic press fully equipped and computerised. The tests were performed at a crosshead speed of 10 mm/min or an overall compression strain-rate of 10^{-3} /sec.

A series of photographs of crushing modes taken during various stages of the collapse process for each specimen are shown in Figs. 2 and 5, respectively, along with terminal patterns of the crushed components; are Figs. 3 and 6.

Load/shell shortening (displacement) curves during the crushing process were automatically recorded by an autographic recorder and plotted in Figs. 4 and 7 for the two materials A and B, respectively. The values of the initial peak load, P_{max} and the energy absorbed W, as well as the mean post-crushing load P (defined as the ratio of energy absorbed to the total shell shortening), are also recorded and tabulated in Table 1.

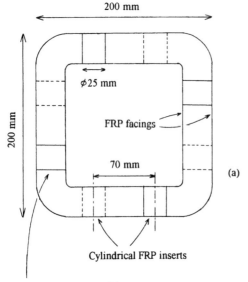

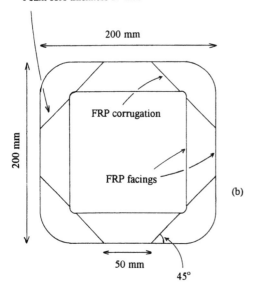

Figure 1. Cross-section of the square composite sandwich tubular components with (a) "Tubular", (b) "Corrugated" core system.

Table 1. Crashworthy characteristics

	Material A	Material B
Initial peak load, P_{max} (kN)	249.1	294.6
Mean post-crushing load, P (kN)	186.2	215.0
Energy absorbed, W (kJ)	33.4	39.1
Effective crush-distance (mm)	184.0	185.0

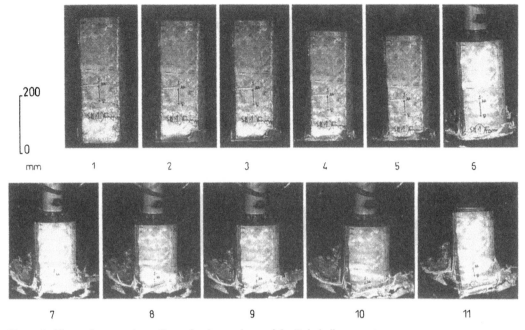

Figure 2. Views of progressive collapse for the specimen of the "tubular" core system.

3 RESULTS AND DISCUSSION

3.1 *Deformation modes*

The characteristic mode of failure for both materials tested shown in Figs. 2 and 5, respectively, is the "end-crushing" mode, characterised by progressive crushing in the form of "mushrooming" failure. In both cases initial collapse occurred at the lower end of the shell and failure progressively spreaded along its height, showing high energy efficiency with load advance. However, after a certain amount of deformation, for the square sections of the "tubular" system the "end-crushing" mode was followed by a rapid propagation of a crack at one corner, leading finally to the entire failure of the tube, see Fig. 2, whilst in the case of the "corrugated" configuration it was followed by a crack propagating along the shell circumference in a position just lower than the top end, see 9-11 in Fig. 5, resulting in the collapse of the tubular component.

During the elastic deformation of the shell the load rises at a steady rate to a peak value, P_{max}, see Figs. 4 and 7, followed by a rapid drop. At this stage, cracks form at the four corners of the shell and propagate upwards along the the tube axis, splitting the side wall into two continuous fronds which spread radially outwards and inwards in the form of a "mushrooming" failure. As deformation proceeds, the externally formed fronds curl upwards with the simultaneous development of a number of axial splits, due to the developed tension in the circumferential direction of the shell, followed by

Figure 3. Terminal views of the collapsed specimen of the "tubular" core system.

splaying of material strips. The length of the splits probably derive the effective column length of the material strips undergoing loading. For both materials, axial tears were not apparent in the internal fronds, which

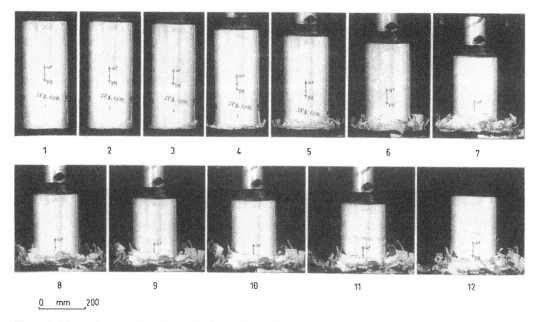

Figure 5. Views of progressive collapse for the specimen of the "corrugated" core system.

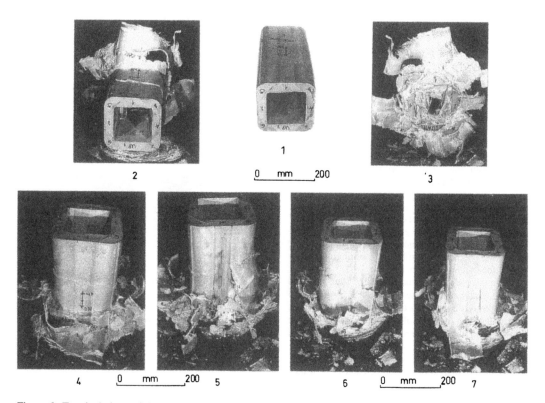

Figure 6. Terminal views of the collapsed specimen of the "corrugated" core system.

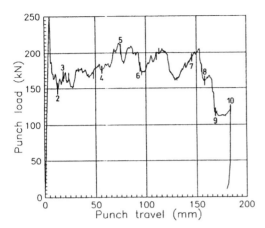

Figure 4. Load / deflection characteristics of axially compressed shell of the "tubular" core system (numbers 2-10 refer to the photographs of Fig. 2).

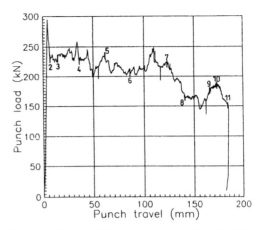

Figure 7. Load / deflection characteristics of axially compressed shell of the "corrugated" core system (numbers 2-11 refer to the photographs of Fig. 5).

were more continuous than their external counterparts. The post-crushing region of the curves is characterised by a highly serrated type, due to the microfracture mechanisms developed inside the hybrid composite material, and starts oscillating about a constant value of the mean post-crushing load. After an amount of deformation of about 130 mm, when corner collapse or circumferential cracking developed, the crushing load reduces at very low levels.

3.2 *Energy absorption*

The energy absorbing mechanisms, encountered to the collapse modes observed, are quite similar to those predicted when crushing fibre-reinforced composite tubes, see Ref. [1]. However, the collapse response of the "hybrid" structures examined is extremely complex, with a variety of macro- and microfailures contributing to the overall energy absorption. The main sources of energy absorption may be classified as:

- Progressive folding of face plates and corrugation.
- Extensive brittle microfragmentation of fibreglass faceplates and corrugation.
- Internal cracking of composite componets, localised fracture of fibres and matrix, splaying of fibres, delamination cracks, flexural damage of external and internal fronds, axial laminate tearing, etc.
- Extensive cracking of the rigid polyurethane foam core material (plastic collapse and pulverisation).
- Frictional resistance to all contact surfaces, internal and external (platen/fronds, platen/core, face plates/core).

Analytical and numerical models, pertaining to predicting the energy absorbing characteristics of these "hybrid" structures, are under investigation.

4 CONCLUSIONS

Hollow square structures, fabricated from composite sandwich panels tied with internal fire reinforced plastic cylindrical tubes or corrugations, have been tested under quasi-static axial compression. Based on the preliminary experimental observations made, some concluding remarks may be drawn:

(a) In general, the failure modes observed throughout the tests seem to be greatly affected by the structure of the hybrid component, the arrangement of fibres, the properties of matrix and fibres of the composite material, the properties of foam and the stacking/fabrication sequences of the designed material.

(b) By positioning fibre reinforced plastic elements within the cores of sandwich panels, mechanisms for controlling the failure loads and, hence, the energy absorption capability of a structure can be incorporated. The principal function of the internal fibre reinforced plastic elements is to provide a mechanical tie between opposing facings. This arrangement enhances the mechanical properties of the sandwich panel, particularly with respect to shear stiffness and strength.

(c) The "corrugated" tie system results in wide continuous areas of contact between the internal structure and the face plates and, subsequently, in a very strong integral design. Conversely, the "tubular" tie system relies on much more localised contacts. This difference, regarding the crashworthy behaviour of the two hybrid structural systems is indicated in Table 1, by comparing the initial peak and the mean post-crushing loads, as well as the energy absorbed.

5 ACKNOWLEDGMENTS

The reported results are part of the experimental, analytical and numerical work on static and dynamic axial collapse of composite rail components related to the BRITE -EURAM BE96-3027 Project "Hybrid composite structures for crashworthy bodyshells and safe transportation structures (HYCOTRANS)". The contribution of the other partners (Antony, Patrick & Murta Exportacao Lda, CETMA Consortium, Costaferroviaria SpA, D'Appolonia SpA, Ifor Williams Trailers Ltd, IKV Aachen, Irizar S. Coop. and University of Perugia) is acknowledged. We are also grateful to Dr. Hans von den Driesch of the EU for his permission of publishing this work.

REFERENCES

Mamalis A.G., Manolakos D.E., Demosthenous G.A. and Ioannidis M.B. "Crash-worthiness of Composite Thin-Walled Structural Components", Technomic Publishing Co, New York, USA 1998.

Mamalis A.G., Robinson M., Manolakos D.E., Demosthenous G.A., Ioannidis M.B. and Carruthers J. "Crashworthy capability of composite material structures", Composite Structures, Vol. 37, pp. 109/134, 1997.

Barbagelata A., Moro E., Mamalis A.G., Manolakos D.E., Robinson A.M. and Walters A.E.D. "Hybrid composite structures for crashworthy safe transportation systems", Proc. 30th ISATA Conference on Crash Behaviour of Lightweight Materials and Structures, Florence, Italy, June 1997, pp. 657/664.

Mamalis A.G., Manolakos D.E., Ioannidis M.B. and Kostazos P.K. "Finite element modelling of the crush zone of fibreglass composite thin-walled tubes subjected to static and dynamic axial loading", Proc. 5th International Conference on Composite Engineering (ICCE/5), Las Vegas, Nevada, USA, July 1998.

Experimental Techniques and Design In Composite Materials 4, Found (Ed.)
© *2002 Swets & Zeitlinger, Lisse, ISBN 90 5809 370 0*

Design of composite vehicle end structures in railway rolling stock

S. Ingleton, M.S. Found & A.M. Robinson
Advanced Railway Research Centre, University of Sheffield, UK

ABSTRACT: This paper discusses the use and design of composite vehicle end structures in railway rolling stock. The paper describes the research being undertaken at ARRC to design and develop such structures for application in the UK heavy rail markets and examines the potential benefits in using such materials. Research into the existing applications of composite end structures are presented and the suitability of the applications of these findings for future developments examined. Key challenges that designers will face in producing composite end structures capable of satisfying current industry criteria are discussed along with the major technical requirements that these composite end structures will have to meet if they are to feature in such applications in the future.

1 INTRODUCTION

During recent years the UK rail industry has been subjected to a number of major changes both in its organizational structure through privatization and also with the introduction of new technological developments. One of the most significant technological developments which has taken place within the rail industry has being the improvement of vehicle safety and in particular, protection to vehicle occupants during collisions.

Until recent new rolling stock introduction, passenger carrying trains running on the Railtrack network were not designed for structural collapse in collision situations, i.e. the trains were not 'Crashworthy'. As a result of work undertaken by British Rail Research new standards GM/RT2100 (1997) were introduced which were to specify not only strength and fatigue requirements of rail vehicle bodies but also their performance regarding controlled structural collapse where peak force levels, deformation and minimum energy levels are stipulated.

The introduction of these standards provided engineers with a new set of challenges which would require fundamental novel design concepts to be employed in order that bodyshell structures would be able to provide structural integrity and yet collapse in a predictable manner.

In meeting the challenges of the new standards, engineers have recently successfully developed and tested vehicle end structures providing high energy absorption capability in traditional materials such as carbon steels and aluminium. The use of composite materials in the design of energy absorbing vehicle end structures has however until recently received little attention or investigation despite widespread use of composite materials in many related applications throughout the world.

The aims of the research being undertaken at the Advanced Railway Research Centre (ARRC) are to design and develop a crashworthy composite vehicle end structure principally to satisfy the standards required in the UK rail environment but also focusing on the engineering and development of techniques suitable for wider use.

The end structures of vehicles are without doubt one of the most complex elements in modern train design and in order to incorporate composites safely into these structures the research will explore a number of key areas, namely:

- Design methodology
- Approach to analysis
- Joint configuration and failure modes
- Energy absorption mechanisms
- Material selection criteria
- Integration of the end structures into the overall bodyshell concept

Vehicle end structures are a vital element in the operation of all new rolling stock and are required to perform a variety of functions ranging from providing structural integrity as part of the overall vehicle body to protection of occupants in collisions and derailments. The end structures are also required to provide many aspects of train operation including provision of safe passage between vehicles, access/egress and aerodynamic styling to name but a few and the research will also need to incorporate these requirements if effective design solutions are to be produced.

Although composite structures are used widely in many industries such as the aircraft and automotive, it is only recently that use of composites for structural applications in the rail industry has started to receive the attention it deserves.

Traditionally in the United Kingdom (UK) rail industry composites have been limited to use for vehicle interiors where these materials have been used extensively for a number of years in applications such as floor, ceiling panels and partitions. In contrast to this, the use of composites for the main bodyshell structural applications, where a high degree of structural integrity is required, has remained relatively unexplored. If this is the case why should the rail industry be interested in composite end structures?

2 BENEFITS OF COMPOSITE END STRUCTURES

There are a number of reasons why the use of composite end structures can be of particular benefit to the industry which can be summarized into three main categories:

- Strength and weight
- Aerodynamics and styling
- Integration of interior systems

2.1 Strength and weight

The high strength to weight ratio of composites, coupled with the capability to optimize strength where required, provides engineers with an opportunity to produce efficient lightweight end structures, particularly as these structures are subject to both severe and varied load cases which often result in current structures being excessively heavy.

Weight, as with most transport related products, is a key criteria and the rail industry is no exception. As rolling stock manufacturers respond to the new private markets, with the emphasis on providing operators with vehicles with a much greater operational flexibility, new rolling stock is becoming heavier. As a consequence of heavier vehicles, energy consumption increases with the heavier axle loads reducing both track life and quality and, as such, weight savings provided by composite end structures provide a whole range of benefits.

2.2 Aerodynamics and styling

Currently the majority of vehicles in the UK provide aerodynamics and styling by superficially cladding the main load bearing structures which, due to the severity of the applied loading, are less irregular and of simple shape to aid production. Composite structures can combine these two requirements, eliminating the need to produce and match two rather large and complex shapes, and also provide an opportunity to generate a more distinctive and artistic approach to end structures. Certainly, with high speed train developments, the styling of the end structures is becoming increasingly important, not just aerodynamically but also visually, as can be seen with the end styling of the trains shown in Figures 1 and 2. Using composites provides designers

Figure 1. Shinkansen high speed train (Japan).

Figure 2. Artists impression of new high speed trains for operation in UK.

with the opportunity to produce vehicle end structures which match the visual inspiration of the industrial designer, something currently rarely achieved but which with composites becomes more feasible.

2.3 Integration of interior and exterior systems

The ability to integrate a number of systems into one single composite element is one of the main areas where composites provide a significant opportunity to remove cost compared to current practice. Currently designs of vehicle end structures are developed and built on a individual system basis, i.e. structures, interiors, electrical installation, ventilation ducts etc.

Composite structures provide an opportunity with their flexibility to accommodate the manufacturing costs from several of these functions into one element, combining the various functions as illustrated in Figure 3.

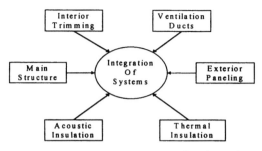

Figure 3. Integration of system functions into composite end structures.

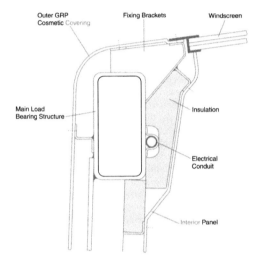

Figure 4. Section through cab corner pillar using conventional design.

An example of the integration of these system functions is to combine the structural strength, aerodynamic, acoustic and thermal insulation requirements into one composite component, eliminating several separate design and manufacturing operations.

This principle is illustrated in Figures 4 and 5 which shows a cross section through a conventional design for a cab corner pillar and a comparable design of the same section in composite materials. As can be seen by comparing the two types of section, the composite design has a number of advantages over current practice in that it offers a reduction in the number of components, provides inherent acoustic and thermal insulation, and the inner and outer laminates of the composite, as well as providing structural strength. It also performs the function of the outer cosmetic covering and interior panel.

When evaluating the cost of producing composite end structures against current practice, the cost of the composite structure should not be simply compared against

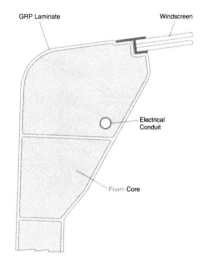

Figure 5. Section through cab corner pillar using composite design.

the cost of an existing replacement in steel or aluminium, but should be measured against the total cost of producing a vehicle end.

This comparison should also include integration of the other systems, such as the outer cosmetic covering, insulation and interior paneling, to provide a true comparison of the existing cost of producing end structures against those in composite.

If integration of systems into the composite structures is to fulfill its potential, then coordinating the design of several systems of produce an overall effective solution with optimum manufacture will be an essential requirement and a new design methodology may need to be developed to incorporate these aspects effectively.

3 USE OF COMPOSITES IN CRASHWORTHY STRUCTURES

At the moment in the UK rail industry no rolling stock incorporates composite structures which are designed specifically to be crashworthy. An attempt to use the energy absorption capabilities of composite structure was however investigated by Scholes & Lewis (1992) as part of collision tests commissioned by the Office for Research and Experiments (ORE), the research arm of the International Union of Railways (UIC). The tests examined the feasibility of a proposed structural design philosophy to examine the crashworthiness of the structure and provide verification of recommended progressive collapse and energy absorption characteristics.

As part of the testing of the crashworthy structures two GRP energy absorption devices were used, one mounted at each buffer position and were designed to collapse at 800 kN (400 kN each). The rear of the tubes were mounted onto the vehicle structure and the front of the tubes fitted with a ribbed anti climb buffer or plate,

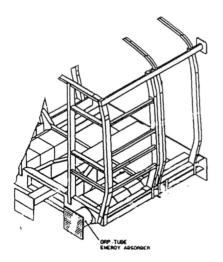

Figure 7. Inter city 125 high speed train composite cab from Robinson & Carruthers (1995).

Figure 6. Cab structure used for laboratory crush test incorporating GRP energy absorbing tubes from Scholes & Lewis (1992).

the purpose of which was to provide resistance to the vertical forces also generated on impact, the ribbed device ensuring that the absorption devices remain in contact during collapse and do not separate.

The GRP energy device used in the test can be seen in Figure 6 along with the basic structural design adopted

The energy absorbing GRP tubes started to collapse in the test at 5% above there combined collapse load of 800 kN which in itself is not significant, but the manner of failure was, as the expected energy absorption characteristics of the tubes were some 50% less than predicted. Scholes et al. (1992) reported that the failure mechanisms of the tubes were due to gross shear failure of the resin matrix causing large fragments to break away rather than de laminate and as a result much of the inherent energy capacity of the elements was lost.

Upon summary of the tests it was recognized by Scholes and Lewis that further work was required in the area of inexpensive, easily replaceable, energy absorption devices, and that the GRP tubes provided poor results for the designs used in this particular test.

4 EXISTING DESIGNS OF COMPOSITE VEHICLE END STRUCTURES

4.1 *Inter City 125 high speed train composite cab structure*

Nearly all rolling stock cabs designed and built in the UK to date have been based on a construction of a load bearing metallic structure, usually steel, surrounded by an outer cosmetic covering of GRP and, even today, this is still the traditional method employed by rolling stock

manufacturers. Use of composite structures has however been successfully introduced into rolling stock cabs in the UK, as early as 1977, in the Inter City 125 High Speed Train cabs as shown in Figure 7.

The cab front of the high speed train was produced from a 50 mm sandwich construction with the outer and inner skins being made of glass reinforced polyester, between which existed a foam core of polyurethane as described by Nock (1980).

The power car of the HST was the leading vehicle and as with all trains protecting the driver from the noise emanating from the diesel engines and wheel/rail interface was an essential requirement. In order to determine the noise performance of the cab environment a prototype train, including the composite cab, was produced and running trials were performed to assess among others, aspects of the train the cab environment. The tests revealed that at higher speeds, noise levels were unacceptable in the cab as a result of which several changes were made to the train generally, and more specifically related to the cab, the entire floor structure was completely redesigned from the prototype. The new floor construction consisted of several acoustical and structurally de coupled layers as well as acoustically absorbent materials. The construction of the floor of the cab can be seen in Figure 8.

It should also be noted that the production vehicle had air conditioning ducts incorporated in the floor as climatic control of the cab was also seen as an essential requirement of producing a suitable cab environment for the driver, particularly as at speeds of 125 mph, where it was not possible to open windows for natural ventilation.

The HST was designed and built prior to the crashworthy standards being introduced which came in after the Clapham rail accident in 1988 and were not required to meet the severe proof and collapse criteria required in modern rolling stock designed to Railtrack Group Standards GM/RT2100 (1997). Whilst the acoustic, thermal and impact properties of the construction were

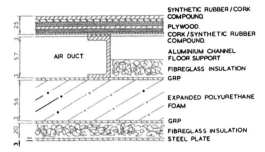

Figure 8. Inter city 125 high speed train composite cab floor construction of production vehicle from Nock (1980).

Figure 9. Inter city 125 high speed train composite cab with collision damage. (Picture courtesy of Interfleet Technology).

Figure 10. Inter city 125 high speed train composite cab. (Picture courtesy of Interfleet Technology).

well established, the crash performance was not a key function of the design and, as such, little or no material exists on the structures collision behaviour.

The HST cab was the first composite structure designed to withstand missile resistance on the BR network and the specification imposed for the forward facing surfaces of the cab are the same as those required on modern rolling stock today as described by Brown (1997).

Modern rolling stock must withstand penetration into the vehicle of a hollow section steel cube, weighing 0.9 kg, travelling at twice the velocity of the vehicles maximum operating speed, in the case of the High Speed Train operating at 125 mph requiring resistance to impact penetration of the steel cube at 110m/s. For the HST the impact resistance testing was carried out by impacting the centre of a panel 1 metre square, the sandwich panels being produced from laminates using unidirectional roving plies laid alternatively at 90 degrees to each other with an injected foam core, as described by Brown et al. (1997).

Figure 9 depicts a picture of a HST cab involved in a collision. As result of the extensive damage to the cab it was replaced rather than repaired with a new cab as shown in Figure 10.

The collision of the HST composite cab highlights some of the difficulties in designing composite structures for collision purposes, where conventionally a certain degree of structural integrity is still required when subject to large scale deformation. As can be seen by the illustration the top area of the structure has been completely separated from the lower portion below the window area, with the high level shearing action causing gross failure of the rear main cab section, exposing the foam cores.

5 MODERN APPLICATIONS OF COMPOSITE STRUCTURAL ENDS

Alberto, Issenmann & Kalbermatten (1991) describe the development of the drivers cab of the new locomotive 2000 series for the Swiss Federal Railways specifically designed to reduce the weight of the locomotive and to introduce fibre reinforced plastics (FRP) into the design. The use of FRP was important in designing the vehicle cab providing opportunities to engineer styled, smooth and attractive shapes for the locomotive, using the full three-dimensional manufacturing capabilities of the material, provide opportunities to optimize aerodynamic requirements and to replace components previously produced in steel with FRP.

For the design two types of construction were initially considered, one with a very lightweight steel structure with a superimposed lightweight glass fibre reinforced laminate and the second a self supporting sandwich structure. The findings of the study are significant in that they revealed a number of key reasons why the superimposed version could not be used when compared to the sandwich structure, namely:

- Little or no weight saving
- Concern over integrating windows and other items where a flush finish is required
- Lack of rigidity and subsequent capability to withstand pressure pulses generated by passing trains
- Poor resistance to missile penetration of small objects
- Requirement for internal sound and thermal insulation still required

Based on the above findings it is not surprising that the self-supporting sandwich construction was selected and produced subsequent weight savings of 1000 kg per locomotive (500 kg per cab).

5.1 Sandwich structure construction

The type of sandwich construction used by Alberto et al. (1991). in the locomotive 2000 consisted of glass fibre reinforced modified polyester resin encapsulating a PVC foam. The sandwich construction is shown in Figure 11 and consists of three layers of glass fibre all of a different thickness and two layers of PVC foam core material. The central trapping layer considerably increased the resistance to missile penetration.

This type of sandwich replaces the 2mm of steel skin attached to conventional steel frame and also the steel frame itself.

When selecting the material for the composite cab Alberto et al. selected a rigid damage tolerant core material of PVC foam to allow expansion due to temperature effects to take place, as the range of temperatures from the inside of an air conditioned cab to the outside skin temperature could be considerable, this temperature differential is an important aspect for the design of the composite structure for material selection.

5.2 Assembly of the cab

Due to the size of the cab of the locomotive Alberto et al. split the design of the sandwich construction into four major parts to aid manufacture, handling and storage, with the final components brought together as shown in Figure 12, with the four individual parts bonded together using a structural adhesive with a special joint arrangement as shown in Figure 13. The part used for the roof section appears to be about two thirds the thickness of the front section described earlier in Figure 11, with the joint configuration connecting the two components providing a large surface area for the adhesive to form an effective joint between the two components.

5.3 Manufacture

Vacuum injection was selected by Alberto et al. (1991) as the manufacturing process, as opposed to a hand lay method, mainly due to the large quantity of cabs required, but also due to several other factors, namely:

- Improved dimensional control of components

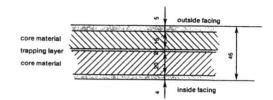

Figure 11. Section of self supporting FRP sandwich construction from Alberto, Issenmann & Kalbermatten (1991).

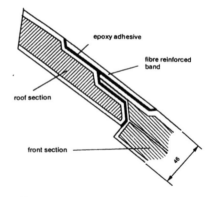

Figure 12. Assembly of cab Locomotive from Alberto, Issenmann & Kalbermatten (1991).

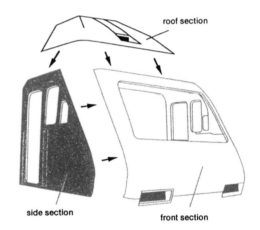

Figure 13. Structural bonded joint design of cab front and roof from Alberto, Issenmann & Kalbermatten (1991).

- Smooth surfaces on both sides
- Repeatability of the components
- Economic production
- Improved Health and Safety to Production Staff

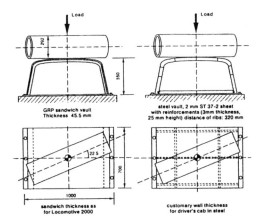

Figure 14. Steel and sandwich panel crush from Alberto, Issenmann & Kalbermatten (1991).

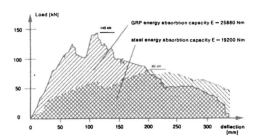

Figure 15. Energy absorption characteristics of the steel and sandwich test panels from Alberto, Issenmann & Kalbermatten (1991).

The sandwich panels were produced in a mould with the precut dry glass fabrics, foam cores and inserts placed between a male and female tool arrangement. Resin was subsequently pumped into the joint and dispersed throughout the mould by means of the vacuum process.

5.4 Impact testing of the sandwich construction

Impact tests on the sandwich panels of 80 cm x 80 cm with a 1 kg impact mass, to UIC standards, with the sandwich section shown in Figure 11, withstand an impact at 280 km/h. For the UK, Railtrack Group Standard GM/RT2100 (1997) states that similar forward facing structures must withstand an impact of twice the maximum speed of the vehicle. Currently, for the UK, the maximum speed of operation is 200 km/h soon to be increased for the West Coast mainline upgrade to 225 km/h, requiring composite structures to withstand an impact resistance of a 0.9 kg mass at 400 km/h and 450 km/h respectively.

Impact resistance is an important aspect of the cab design as this provides protection to the driver from projectiles such as track ballast disturbed by the slip steam of passing trains and, more seriously, from stones and bricks deliberately thrown at trains by vandals.

The results of the tests performed by Alberto et al. (1991) revealed that at 280 km/h the damage remained localized and penetration of the object was only partial. Interestingly when a rigid core material was used a larger area of the sandwich panel was destroyed.

5.5 Collision testing of the sandwich construction

Alberto et al. (1991) also performed tests on the sandwich construction to compare the collision performance of the sandwich construction with that of a steel construction. Two test pieces were constructed, one in each material, and the energy absorption capacity of the materials measured when crushed. The testing of the samples and the resultant energy absorption are shown in Figures 14 and 15 respectively.

The samples were compressed under load to 38% of there original height with the sandwich construction absorbing 25% more energy than the steel counterpart.

The tests performed by Alberto et al. did not discuss the failure mechanisms of the sandwich panel or the force displacement characteristics, but from the results of the test there are several aspects which provide useful criteria for the behaviour under collision of composite structures.

The sandwich structure peak collapse force shown in Figure 15 was almost twice that of the steel structure, and exhibited a decreasing force level as deflection increased, compared to the steel sample which increased gradually. This suggests that the sandwich structure started to delaminate or fracture when compared to the steel sample which increased its resistance to the applied load.

Peak loads are an important aspect of the Crashworthiness philosophy in the UK, and are specified in the Railtrack Group Standard GM/RT2100 (1991), as they are related directly to passenger safety as a function of acceleration during train collisions. The control of peak forces for the design of the crashworthy composite structure is an area where methods must be explored to control such behaviour.

For the sandwich construction the force dropped to almost zero where as for the corresponding deflection the steel sample retained a load approaching its average collapse force.

The composite cab of the locomotive 2000 in passenger service can be seen in Figure 16.

5.6 Suitability of findings for crashworthy cab

The sandwich construction cab produced for the locomotive 2000 produced very interesting information about the collision behaviour of composite structures and a number of aspects for consideration.

Figure 16. Cab end nosing of locomotive 2000 in passenger service (Picture courtesy of Alusuisse).

There are however several elements not covered or inherent in the design of the cab which have a significant effect on the production of crashworthy structures.

The cab was not designed specifically to be crashworthy, and the results of the collision test and the manner in which the load was applied have no resemblance to the crashworthy standards adopted for the UK by Railtrack. The design of the cab does not explore the controlled collapse of the entire cab structure but merely provides, by way of a test piece, a benchmark from which the performance can be compared to that of a steel test piece.

Controlling peak forces are not discussed or set as a design parameter, aspects which along with the very high proof load required for the UK are dictated in GM/RT2100 (1997).

Collapse mechanisms and the behaviour of joints are not specifically discussed in the work of Alberto et al. (1991) but overall provide a clear indication into areas where some of the major issues lie with producing these type of structures.

5.7 *Structural driving cab Of C20 Stockholm Metro car*

For the design of the C20 Stockholm Metro Car great emphasis was placed on the styling of the drivers cab and, in order to deliver a design which satisfied the aspirations of the customer, the decision was made to produce the driving structures in composite materials. Use of composites to produce stylish shapes is not new, even to the rail industry, as traditionally composites have been used to provide aesthetic front end shapes as superficial cladding to a substantial but hidden metallic load bearing structure for many years.

The main difference of the C20 against the 'traditional method' of manufacture was that the composite structure was to provide not only appearance characteristics, but would be also capable of withstanding a wide range of structural loads applied locally to the vehicle end and contribute to the overall stiffness performance of the vehicle structure as described by Werne (1997).

Werne et al. describes the design of the structure employed for the C20 cab which consisted of a GRP/balsa sandwich construction with the skins being produced from knitted multiaxial fabrics in a polyester matrix. The cab was 2m long and 3m high and produced as a one piece hand laminate, which in turn was bolted to the main vehicle structure as a modular unit, i.e. 'bolt on cab'.

From the resulting design of the cab Werne et al. concluded that there was much to be gained by varying the laminate lay up over the structure to satisfy particular requirements, with the more heavily loaded forward facing structural areas using thicker laminates reinforced predominantly in one direction with a 50 mm core, as opposed to the roof and side structures which employed thinner laminates with quasi-isotropic reinforcement and a 25 mm core.

The cab design included in-built thermal insulation thus removing the requirement from the interior trimming and made use of 'Bighead' fasteners laminated into the inner sandwich skin to mount major pieces of interior equipment such as air conditioning units and drivers door actuating mechanisms.

5.8 *Comments on findings of C20 with respect to UK heavy rail*

One of the key areas of the C20 cab is its capability to withstand the structural loads placed upon the end structures as well as provide styling. In the UK the load cases applied to the vehicle end structures for the heavy rail vehicles are very severe and are classified in terms of proof, fatigue and collapse requirements in GM/RT2100 (1997). The C20 load cases by comparison are relatively small, few in number and no collapse requirements in case of collision are specified.

The work undertaken on the C20 does provide an indication that the proof load design within structures can be met with careful design but provides no evidence of collapse characteristics of composites when loaded beyond these levels.

6 CHALLENGES WITH USE OF COMPOSITE MATERIALS FOR END STRUCTURES

Using composite materials for crashworthy vehicle end structures provides a tremendous challenge for the future, but what are the real challenges that lie ahead?

In general fibre reinforced plastics do not exhibit a ductile failure mode when compared to their steel counterparts, but tend to exhibit a brittle mode of failure with high peak forces and catastrophic degradation of the laminates. For UK rolling stock the design requirements for proof loading and collision are some of the most stringent in the world, and, as result to date, the preferred material for such application is low carbon steels. Although composites used in rolling stock cabs provide many advantages and evidence suggests they are capable of meeting the high proof load requirements, meeting the collision performance is quite different. Designing a composite structure to absorb one or two MJ of energy with peak forces below 3000 kN requires composite structures to absorb energy over areas of large deformation. This is the real challenge with the composite structure designs, as the brittle failure modes typically exhibited by fibre reinforced plastics combined with inherent strength, requires these failures under collisions to be controlled by limiting peak force whilst retaining a high degree of structural integrity to protect occupants as gross failure is occurring.

7   TECHNICAL REQUIREMENTS OF DESIGNING COMPOSITE END STRUCTURES

In order to design composite structures for the UK rail network there are a number of technical requirements which the end structures must meet, if they are to be allowed to be used in passenger carrying service, some of these requirements are mandatory regulation, and some inherent requirements due to the nature of the operating environment.

The requirements can be broadly categorized into four main groups, namely:

• Structural
• Environmental
• Production
• Operation

7.1   Structural requirements

There are a number of structural requirements which the end structures must satisfy and these are well defined in Railtrack Group Standard GM/RT2100 (1997) and cover aspects ranging from proof and collision loads to fatigue and missile protection.

The structural requirements are a dominating feature in the design of the vehicle ends, where the severity and variety of loads applied test the structure particularly in terms of proof and collapse requirements. Typical proof loads vary from 1500 kN at the underframe coupler position to 300 kN at the upper cantrail areas at the front of the structure.

Although fatigue is not normally of significance in the design of the structure at the vehicle ends, due to the dominance of proof and collapse loads, it is of greater significance with the use of composites, particularly as the front of trains often receive minor superficial damage which may result in crack propagation being undetected, and this aspect will require assessment, possibly by a risk based approach.

Missile protection is required in the structure to protect the driver from injury caused by missile penetration into the drivers cab, and evidence suggests that composites provide this feature inherently where laminates and foam cores are used. The requirement for missile protection are discussed in detail in section 5.

7.2   Environmental requirements

Fire performance is a key requirement for trains, the details of which are contained within British Standard 6853 (1987). The requirements of this standard are particularly onerous and the standard describes the requirements for spread of flame and toxicity and specific testing which is required in order to demonstrate compliance, this aspect will be of major significance when selecting suitable composite materials.

The composite cab must be able to cope with the demands of climatic change ranging typically from sub zero temperatures of -15 degrees centigrade to elevated temperatures of +30 degrees centigrade.

The composite selected for the construction will be operating in a harsh environment, where it will be subject to an operational diagram of daily cleaning and the chemical agents used in this cleaning process, coupled with use of oils and greases found in maintenance depots, can be particularly aggressive.

Ingress of moisture into the composite structure must also be guarded against, particularly where foam cores are used to prevent frost damage.

The composite structures will also be required to be painted to suit operator livery and withstand removal of graffiti and any chemical agents required to facilitate this removal.

If the material selected can offer recycling capabilities then this would also be a desirable objective, even if not a mandatory requirement, as disposal will eventually be required.

7.3   Production requirements

Although there are no specific Railtrack Group Standards relating to the production of the end structure, there are several considerations which need to be reflected in the design if composite end structures are to succeed.

The design will need to be modular to reflect market changes by providing an adaptable base product with an inbuilt degree of flexibility to suit a range of vehicle applications.

The component itself will require to be cost effective, issues which are discussed earlier in section 2 and which deals with the integration of other systems into manufacture and production. The composite material will require a production process capable of being self supporting, providing repeatability and produced to tolerances which complement those required for interfacing the component to the main bodyshell structure.

There are significant amounts of production work to be undertaken in vehicle end structures and the methodology for producing these ends and the build sequence will require considerable planning at the conceptual design level in order to capitalize on this aspect to minimize labour costs.

As the market for rail vehicles enters a new era with trains being potentially owned by several operators over the life of a vehicle, the operational characteristics become increasingly important. The end structure of a rail vehicle will be typically designed for 30 years and the cost of the component over this period should therefore also be a consideration of the product not just the 'first cost'.

Vehicle structures suffer damage in service ranging from small abrasion to minor and major impacts. In order to reduce the amount of time a vehicle is out of service, the vehicle conceptual design should address the issue of vehicle damage by repair or replacement. Potential repair scenarios of damaged end structures need to be evaluated as an inherent feature from the start of the design process, not as an afterthought when the product has its first incident.

The composite end structures are also driven by the key requirements of geometry and the need to produce effective aerodynamic and stylish end shapes as described in section 2. Furthermore the external shapes are tapered in plan elevation to avoid contact with tunnels, bridges and platforms as the vehicle overthrows on negotiating track curvature.

Finally a key factor in the design of the end structures will be the ergonomic relationship of the train driver within the boundaries of the end structure itself, and the need to cater for the requirements of providing driver visibility of track and signals ahead.

8 CONCLUSIONS

The research work undertaken shows clearly that it is possible to design composite end structures for rolling stock, and there are a number of vehicles across Europe in passenger service which adequately demonstrate this for a range of operating applications.

The requirements to introduce collision performance into the composite designs represents a step change in technical complexity and requires a new approach to provide designs which are capable of meeting this and the stringent technical requirements required for heavy rail passenger service in the United Kingdom.

9 ACKNOWLEDGEMENTS

One of the authors (S.I.) would like to thank the Advanced Railway Research Centre (ARRC) for funding this work.

REFERENCES

Railtrack Railway Group Standard GM/RT2100 1997. *Structural Requirements For Railway Vehicles*.

Scholes A. & Lewis J.H. 1992. Development of crashworthiness for railway vehicle structures. *Proceedings of the Institute of Mechanical Engineers Ordinary Meeting*.

Robinson M. & Carruthers J.J. 1995. Composites make tracks in railway engineering. *Reinforced Plastics*: Vol. 39, No. 11, pp. 20-26.

Nock O.S. 1980. Two miles a minute. *The Story Behind The Conception And Operation Of Britain's High Speed And Advanced Passenger Trains*: pp. 72-83. London: Book Club Associates.

Brown D.E. 1997. The impact of composites on railway rolling stock. *Composites In The Rail Industry Conference Railview Ltd. October 1997*: pp. 1-13.

Cortesi A., Issenmann T. & De Kalbermatten T. 1991. Light nose for fast locomotives. *Airex Composites Manufacturing Special Issue*: pp. 435-442.

Werne D.V. 1997. Structural GRP front for metro car. *Composites In The Rail Industry Conference Railview Ltd. October (1997)*.

British Standard 6853. 1987. Fire precautions in the design and construction of passenger rolling stock. *British Standard Institution*.

Section 4: *Impact*

Experimental Techniques and Design In Composite Materials 4, Found (Ed.)
© 2002 Swets & Zeitlinger, Lisse, ISBN 90 5809 370 0

Static indentation and low velocity impact damage in thin composite laminates

F. Aymerich & P. Priolo
Dipartimento di Ingegneria Meccanica, Università di Cagliari, Piazza d'Armi - 09128 Cagliari - Italy

ABSTRACT: One of the major concerns about the use of composite materials is their sensitivity to low-velocity impacts, likely to occur during manufacture or normal operations. Since impact-induced damage may lead to significant reductions in strength, it is important to have a precise understanding of how the damage develops under different conditions such as static and dynamic loading.

In this study both static indentation-flexure and low-velocity impact tests have been carried out on different configurations of liminates made of graphite/epoxy and graphite/peek composites. The internal damage, investigated by non-destructive techniques, has been related to different mechanical parameters obtained by instrumented testing. The results of impact and static tests have been compared to characterize the influence of impact velocity and the role of individual damage modes in the failure process.

1 INTRODUCTION

One of the most dangerous damage modes in thin composite laminates is that induced by low-velocity out-of-plane loads. This damage consists of delaminations, matrix cracks and fibre fractures, often undetectable by visual inspection, which lead to significant reductions in strength. A clear understanding of the different failure modes arising during impact, as achievable by a detailed analysis of the force-time history, is necessary to develop general models able to describe the response of the laminate to dynamic loads and to predict post-impact properties.

The impact response of composite materials depends on many parameters (impact energy, impactor mass and velocity, specimen geometry etc.). The influence of impactor velocity has often been overlooked in that in low-velocity impacts the contact duration is much longer than the travel time of flexural waves to the boundary of the sample. For this reason many authors have suggested characterizing the impact damage resistance by means of simple static indentation-flexure tests [Lagace et al. 1993, Kwon & Sankar 1993], neglecting the differences in structural response due to dynamic effects. However, whereas in conventional epoxy-based laminates the static indentation behaviour is reported to be similar to the response to low-velocity impacts [Sjoblom et al. 1988, Lee & Zahuta 1991, Wu & Shyu 1993, Kaczmarek & Maison 1994], several experimental investigations point to a significant strain-rate dependence of tougher thermoplastic composites [Gillespie et al. 1987, Smiley & Pipes 1987, Mall et al. 1987], thus

suggesting possible changes in failure modes between static and dynamic loads.

In this study both static indentation-flexure and low-velocity impact tests were carried out on different configurations of laminates made of graphite/epoxy and graphite/peek composites. The internal damage - investigated by means of X-ray and ultrasonic techniques - was related to different mechanical parameters obtained by instrumented testing.

The results of impact and static tests were compared to characterize the influence of impact velocity and the role of individual damage modes in the failure process.

2 MATERIALS AND EXPERIMENTAL PROCEDURE

The impact behaviour of laminates of three different composite materials (graphite/PEEK and two graphite/epoxy systems) was examined.

The first was graphite/PEEK with 63% by volume of AS4 fibres, chosen as being representative of the class of tough thermoplastic composites. The graphite/epoxy systems were Grafil 6KXAS fibres in Fibredux 914 resin and MR50K fibres in LTM26 resin, both with 60% fibre volume fraction.

Two quasi isotropic layups ($[0/\pm 45/90]_{2s}$ and $[45/-45/0/90/90/0/-45/45]_s$ both 2.2 mm thick) were used for the APC2 composite, while the stacking sequences of the epoxy materials were $[\pm 45/0/90]_{2s}$ (thickness = 2 mm) for 6KXAS/914 and $[0_4/45/0_4/-45/0_4]$ (thickness = 2.5 mm) for MR50K/LTM26.

Both static indentation-flexure and low-velocity impact tests were conducted utilizing a servohydraulic testing machine and an instrumented drop-weight tower respectively [Aymerich et al. 1996]. The same support fixture was used for the two testing conditions, with the specimens clamped between two 70 mm steel rings and loaded by an indenter with a hemispherical tup of 12.5 mm. Impact energy was varied over the range 2-12 J changing drop height and/or impactor mass; the corresponding impact velocities ranged between 1.3 and 4.8 m/s. Static tests were conducted to predetermined indentor displacements at constant velocity of 0.025 mm/s.

The contact force between impactor and specimen was measured with a semiconductor strain-gauge bridge bonded to the tup. Determination of the real load responsible for the global deformation of the specimen would require an accurate knowledge of the dynamics of the impactor-specimen system. Given the small specimen to impactor mass ratio, the vibrational aspects of the impact can be neglected and the contact force can be taken as a measure of the real force acting on the laminate.

In impact tests the displacement was evaluated by double integration of the contact force signal, and included local indentation in the contact region, while the kinetic energy lost by the impactor was obtained from the impact and rebound velocities; if the energy absorptions in the test rig are reduced, the energy lost by the falling mass during impact is mainly absorbed by the specimen as fracture work and plastic deformation. For static tests the indentor displacement was measured with an LVDT and the energy-time history calculated by integrating the force vs displacement curve.

Since the morphology of impact damage is very complex and no single parameter is sufficient to characterize completely the internal state, several techniques were used to examine the composite samples after testing.

Damaged specimens were X-rayed after infiltration of a radio-opaque dye (a zinc iodide solution) and ultrasonically C-scanned in a water tank using a purpose-built scanning system. The ultrasonic system is complete with a pulser receiver, a digital oscilloscope and a PC acting as a controller. During ultrasonic inspection the specimen is scanned in pulse-echo mode by a 50 MHz focussed probe and the entire ultrasonic wave, including the front and back surface reflections, is digitized at predetermined acquisition points. The acquired waves are then processed in order to obtain ply-by-ply images (time of flight or amplitude C-scans) of the delamination at the desired depth. The total delamination area induced by impact is calculated by summing the delamination areas at single interfaces.

A typical C-scan image consists of a 300 by 300 array and the required acquisition time is about 1 hour. All the samples were scanned from the two sides and the information was recombined to a single volumetric image.

As for fibre fracture, the length of fibre fracture paths at the backface, measured perpendicularly to the fibre orientation, and determined by light microscopy observations or by X-ray examination, was used as indicator of this damage mode.

3 KEY PARAMETERS OF IMPACT TESTS

A great deal of information on the impact event and related damage modes can be gained by analysing the data acquired through instrumented impact testing. The most common forms of representation of impact data are the contact force/energy versus time and contact force versus impactor displacement plots and different parameters drawn from these curves have been used [Ghasemi Nejhad & Parvizi-Majidi 1990, Strait et al. 1992, Teti et al. 1985] to characterize the impact behaviour of structural materials. They range from force/displacement or stiffness indicators to energy-based parameters but no single indicator is usually considered sufficient to classify the different materials in terms of impact resistance or to identify the real sequence of fracture modes.

Typical curves for a 5.2 J impact on a quasi-isotropic $[0/\pm45/90]_{2s}$ graphite/peek composite laminate are shown in Figure 1. Based on these curves, various char-

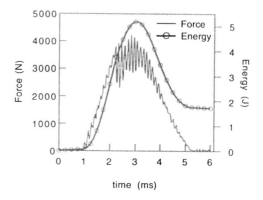

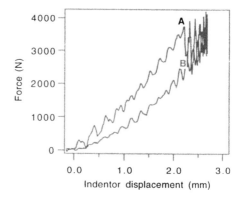

Figure 1. Force/energy-time and force-displacement curve for a 5.2 J impact on a $[0/\pm45/90]_{2s}$ graphite/PEEK laminate.

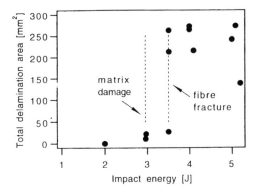

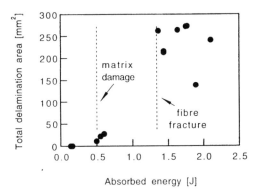

Figure 2. Total delamination area versus impact (top) or absorbed energy (bottom) in $[0/\pm45/90]_{2s}$ graphite/PEEK laminates.

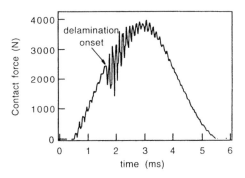

Figure 3. Contact force history for a 4.4 J impact on a $[\pm45/0/90]_{2s}$ graphite/epoxy laminate.

acteristic parameters, some of them associated with single relevant points of the plot, can be defined:

• The impact and absorbed energy, which correspond respectively to the maximum energy imparted to the sample during impact (when no through penetration occurs) and to the energy permanently absorbed by the plate at the end of impact; the latter is mainly dissipated through work of fracture (matrix cracks, delamination and fibre fracture or debonding), plastic deformation and friction work on delaminated interfaces. Both impact and absorbed energy have often been directly correlated to impact damage and are probably the parameters most commonly used to characterize impact performance.

The total delamination area of impacted $[0/\pm45/90]_{2s}$ graphite/peek laminates is plotted versus impact or absorbed energy in Figure 2; the levels corresponding to initial damage (first matrix cracks and delamination around the contact area [Aymerich & Priolo 1998]) or major fibre fracture events are easily identified.

One the other hand, it should be pointed out that with the use of energy-based parameters it is easy, due to the peculiarity of the integrated quantities, to

overlook minor but nonetheless important features of the contact load history, indicative of individual failure modes.

• The incipient damage load (and corresponding energy), indicated in the force-time curve by the first significant discontinuity or by high amplitude oscillations after inertial effects have ceased, which denotes the onset of damage phenomena in the sample. The first failure modes of impacted laminates consist of matrix damage, but matrix cracks cannot be usually detected, due to the minor effect on the global stiffness. In conventional thermoset composites significant delaminations, which frequently occur, can easily be recognized (Figure 3); in thermoplastic composites, on the contrary, initial matrix damage is so limited that its identification is often impossible through simple analysis of the contact load history.

• The maximum contact force, which represents the highest load withstood by the composite panel under impact; the maximum force has been reported [Zhou 1995, Rydin et al. 1995] to increase with impact energy up to a threshold value and then to level off, thus indicating the load-carrying capability of the laminate; beyond this threshold the sample could withstand higher impact energies but only at the expense of significant deterioration in liminate integrity. As a matter of fact, the maximum load is often followed by a sudden load drop (see Figure 1) which denotes the occurrence of major fibre fractures which can severely impair total stiffness. In thermoplastic laminates this is frequently the only damage form clearly identifiable through the examination of the force history.

As opposed to impact energy, the maximum force has been proposed as the key parameter in impact damage resistance, in view of the good correlation observed [Lagace et al. 1993, Zhou 1995, Jackson et al. 1995] between force and delamination area, at least until perforation occurs at high impact energies. The difference between two characteristic

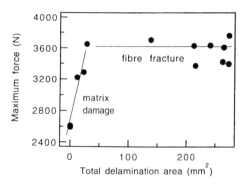

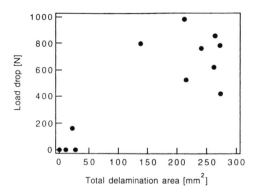

Figure 4. Maximum force versus total delamination area in $[0/\pm45/90]_{2s}$ graphite/PEEK laminates.

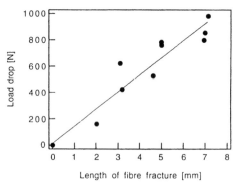

Figure 5. Load drop vs delamination area (top) and length of fibre fracture (bottom) for $[0/\pm45/90]_{2s}$ graphite/PEEK laminates.

regions, the first associated to matrix damage mechanisms and the second to tensile fibre fracture phenomena, is clearly visible in the plot of Figure 4, which shows the maximum force as a function of delamination area for $[0/\pm45/90]_{2s}$ APC2 laminates.

• The amplitude of the load-drop at maximum load (length A-B in Figure 1), caused by unstable fibre fracture phenomena, which can be assumed as a direct measure of the extent of fibre breakage. By plotting the load drop vs delamination area or length of broken fibres (as obtained by X-ray and light microscopy examination) we can notice that this parameter shows no correlation with the extent of matrix damage, but exhibits a selective sensitivity to fibre fracture events (Figure 5) and could thus be used as a reliable indicator of the extent of broken fibres.

It is worth noting that sometimes a sudden decrease in load is observed in correspondence with unstable propagation of delaminations (see Figure 3); this decrease, related to matrix damage mechanisms, does not significantly reduce the load-carrying performance of the laminate and is therefore easily identified, since it occurs at load levels much lower than the maximum attainable load.

• The change in stiffness due to fracture events. All the force-displacement curves present, after a first portion disturbed by inertial oscillation, a non-linear behaviour characterized by an increase in slope, initially related to indentation phenomena and then to membrane effects (typical at large plate deflection) and possibly to friction between tup and specimen. Actually, the first loading part of the curve (at load levels below the onset of serious damage) can be well predicted by finite element analysis, if the proper option of large displacements is introduced. After the load drop associated with major fibre fracture, a significant loss of stiffness is observed, but the presence of high frequency oscillations excited by the sudden release of fracture energy often makes

a reliable measure of the reduced slope difficult. The unloading phase exhibits again a strong nonlinear behaviour, with stiffness largely dependent on the load level and sometimes, at the beginning of the rebound, even greater than the stiffness of the undamaged sample (Figure 1). This strong nonlinearity can be explained by the friction between surfaces created by damage events, especially layers at delaminated interfaces, which tend to open with decreasing loads. In view of these observations, the change in stiffness between the loading and unloading phases, measured at low load levels, might be taken as a possible indicator of impact damage resistance. Figure 6 shows the correlation between the change in stiffness of impacted graphite/peek laminates (evaluated over the range 200-700 N) and the total delamination area or the lenght of broken fibers, selected as parameters representative of impact damage.

4 INFLUENCE OF IMPACT VELOCITY

It is well known that the impactor velocity has a strong influence on the impact behaviour of mechanical components, with responses ranging from localized modes

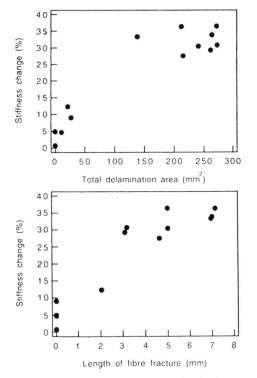

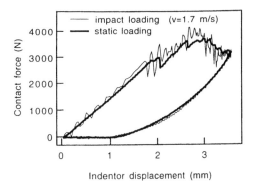

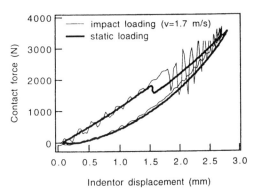

Figure 6. Stiffness changes vs total delamination area (top) and length of fibre fracture (bottom) for impacts on $[0/\pm45/90]_{2s}$ graphite/PEEK laminates.

Figure 7. Force displacement curves for static and impact loaded quasi-isotropic graphite/PEEK (top) and graphite/epoxy (bottom) laminates.

of deformation at ballistic velocities to more global sample deflections at lower velocities; if we focus the attention on low velocity impacts, one of the key factors in the structure's impact response is the rate sensitivity of the material's mechanical properties. As already mentioned, various studies have shown only negligible, if any, influence of impact velocity on the behaviour of impacted graphite/epoxy specimens of different sequences, thus confirming that static indentation-flexure tests can be adopted to characterize their dynamic response.

Interlaminar toughness tests on composites based on viscoplastic matrices such as PEEK, on the other hand, have revealed a strong rate sensitivity, therefore indicating a possible different response between impact and quasi-static loads.

In order to examine the influence of impact velocity, both static indentation and impact tests were carried out on $[45/-45/0/90/90/0/-45/45]_s$ graphite/peek and $[\pm45/0/90]_{2s}$, $[0_4/45/0_4/-45/0_4]$ graphite/epoxy limi-nates. The duration of the contact period ranged from 3 to 8 ms (impactor velocities between 1.7 and 4.8 m/s) for impact tests and from 100 to 150 s, depending on the maximum deflection, for static tests.

The force-displacement curves (Figure 7) show only small differences in structural response between static and impact tests for both graphite/PEEK and graphite/epoxy specimens, with slightly higher stiffnesses associated to higher loading rates.

By plotting absorbed energy versus impact energy, comparable energy absorption values can be observed, for all the materials tested, at the different loading rates (Figure 8); this behaviour suggests that similar failure modes may occur in impacted and statically loaded specimens, and actually graphite/epoxy specimens exhibited the same damage pattern for both static and dynamic loads. If, however, we plot the total delaminated area as a function of impact energy (Figure 9), we notice that, in spite of good agreement between the static and dynamic damage responses of graphite/epoxy samples, a large discrepancy exists for thermoplastic graphite/PEEK laminates, where higher load rates are associated to larger delaminations. The difference in damage extent between a static- and an impact-loaded graphite/PEEK sample - as obtained by X-radiography can be seen in Figure 10.

To explain this behaviour, the lengths of visible broken fibres at the backface of impacted and statically

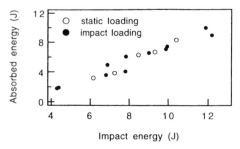

a) [0$_4$/45/0$_4$/-45/0$_4$] graphite/epoxy

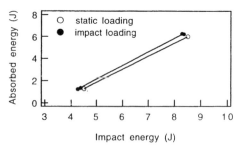

b) [±45/0/90] graphite/epoxy

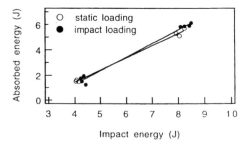

c) [45/-45/0/90/90/0-45/45]$_S$ graphite/PEEK

Figure 8. Absorbed vs impact energy for graphite/PEEK and graphite/epoxy laminates.

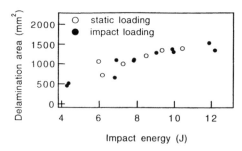

a) [0$_4$/45/0$_4$/-45/0$_4$] graphite/epoxy

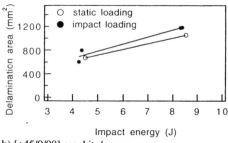

b) [±45/0/90] graphite/epoxy

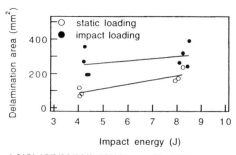

c) [45/-45/0/90/90/0-45/45]$_S$ graphite/PEEK

Figure 9. Total delamination area versus impact energy for graphite/PEEK and graphite/epoxy laminates.

loaded APC2 samples were measured and the averaged values compared in Table 1. From these data it emerges that static tests produce larger amount of broken fibres; it is then possible to deduce that the response of graphite/PEEK laminates shows a distinct rate dependency, with viscoelastic effects dominating at low load rates and thus resulting in failure modes characterized by reduced matrix damage and, as a consequence, significant fibre fracture. At higher impact velocities the time-dependent matrix deformation tends to become negligible and this translates into a change towards matrix controlled failure mechanisms.

Table 1. Length of broken fibres at the backface of APC2 samples

Energy	Static load	Impact load
4.2 J	6.3 mm	1.2 mm
8.2 J	13.0 mm	8.0 mm

This fact confirms yet again that global parameters like absorbed energy are not capable of completely describing impact effects. As these are determined by the local prevalence of one of several competing damage mechanisms, appropriate features such as the

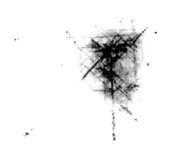

Static load - energy = 4.2 J

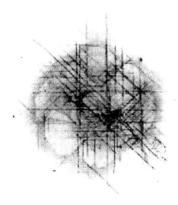

Impact load - energy = 4.2 J

Figure 10. X-radiographs of damaged [45/-45/0/90/ 90/0/-45/45]s graphite/PEEK laminates.

response to matrix delamination and to fibre fracture growth under different deformation rates have to be taken into account.

5 CONCLUSIONS

On the basis of the results of static indentation-flexure and low-velocity impact test conducted on graphite/ epoxy and graphite/PEEK laminates, some general conclusions can be drawn:

- The experimental force-time and force-displacement curves are very useful for identifying the main damage phases of the failure process. Various parameters were defined to characterize the impact behaviour and the damage resistance of the selected materials from different perspectives.
- Impact response, absorbed energy and the damage pattern of impacted graphite/epoxy specimens were similar to those of statically loaded specimens.
- Static and dynamic-loaded graphite/PEEK lami-

nates showed similar energy absorption values but impacted samples exhibited higher delamination areas and lesser extent of fibre fracture, thus indicating a clear influence of the load rate on the damage mode. The simulation of impact response by means of static indentation tests can be misleading as far as damage resistance and tolerance of the laminate is concerned.

- Due to the strain rate dependency, global parameters like impact or absorbed energy do not appear to be able to completely represent the impact behaviour of thermoplastic composites.

   Similarly, widely used damage indicators such as the projected or the total delamination area appear inadequate for properly describing the damage induced by the impact.

REFERENCES

Aymerich F., Marcialis P, Meili S. & Priolo P. 1996. An Instrumented Drop Weight Machine for Low Velocity Impact Testing, *Proc. Structures under Shock and Impact, Udine (Italy), 3-5 July 1996*, N. Jones, C.A. Brebbia & A.J. Watson, Eds., Computational Mechanics Publications, Southampton: 243-253.

Aymerich F., Bucchioni A. & Priolo P. 1998. Impact Behaviour of Quasi-Isotropic Graphite/PEEK Laminates, Proc. Experimental Techniques and Design in Composite Materials, Cagliari, October 1996, *Key Engineering Materials*, 144: 63-73.

Ghasemi Nejhad M.N. & Parvizi-Majidi A. 1990. Impact behaviour and damage tolerance of woven carbon fibre-reinforced thermoplastic composites, *Composites*, 21(2): 155-168.

Gillespie Jr J.W., Carlsson L.A. & Smiley A.J. 1987. Rate-Dependent Mode I Crack Growth mechanism in Graphite/Epoxy and Graphite/PEEK, *Composites Science and Technology*, 28: 1-15.

Jackson W.C., Poe C.C. Jr. 1995. The Use of Impact Force as a Scale Parameter for the Impact Resistance of Composite Laminates, *Journal of Composites, Technology and Research*, 15(4): 282-289.

Kaczamerek H & Maison S. 1994. Comparative Ultrasonic Analysis of Damage in CFRP under Static Indentation and Low Velocity Impact, *Composites Science and Technology*, 51: 11-26.

Kwon Y.S. & Sankar B.V. 1993. Indentation-Flexure and Low-Velocity Impact Damage in Graphite Epoxy Laminates, *Journal of Composites, Technology and Research*, 15(2): 101-111.

Lagace P.A., Williamson J.E., Wilson Tsang P.H., Wolf E. & Thomas S. 1993. A Preliminary Proposition for a Test Method to Measure (Impact) Damage Resistance, *Journal of Reinforced Plastics and Composites*, 12: 584-601.

Lee S.M. & Zahuta P. 1991. Instrumented Impact and Static Indentation of Composites, *Journal of Composites Materials*, 25: 204-222.

Mall S., Law G.E. & Katouzian M. 1987. Loading Rate Effects on Interlaminar Fracture Toughness of a

Thermoplastic Composite, *Journal of Composite Materials*, 21: 569-579.

Rydin R.W., Bushman M.B. & Karbhari V.M. 1995. The Influence of Velocity in Low-Velocity Impact Testing of Composite Using the Drop Weight Impact Tower, *Journal of Reinforced Plastics and Composites*, 14: 113-127.

Sjöblom P.O., Hartness J.T. & Cordell T.M. 1988. On Low-Velocity Impact Testing of Composite Materials, *Journal of Composite Materials*, 22: 30-52.

Smiley A.J. & Pipes R.B. 1987. Rate Effects on Mode I Interlaminar Fracture Toughness in Composite Materials, *Journal of Composite Materials*, 21: 670-687.

Strait L.H., Karasek M.L. & Amateau M.F. 1992. Effects of Seawater Immersion on the Impact Resistance of Glass Fiber Reinforced Composites, *Journal of Composite Materials*, 26: 2118-2133.

Teti R., Langella F., Crivelli Visconti I. & Caprino G. 1985. Impact Response of Carbon Cloth Reinforced Composites, *Proc. 5th International Conference on Composite Materials, San Diego, 1985*, W.C. Harrigan Jr, J. Strife, A.K. Dhingra, Eds.: 373-381.

Wu E. & Shyu K. 1993. Response of Composite Laminates to Contact Loads and Relationship to Low-Velocity Impact, *Journal of Composite Materials*, 27(15): 1443-1464.

Zhou G. 1995. Prediction of impact damage thresholds of glass fibre reinforced laminates, *Composite Structures*, 31: 185-193.

Experimental Techniques and Design In Composite Materials 4, Found (Ed.)
© *2002 Swets & Zeitlinger, Lisse, ISBN 90 5809 370 0*

Perforation of thin plain and stiffened CFRP panels

M.S. Found, I.C. Howard & A.P. Paran
Department of Mechanical Engineering, University of Sheffield, UK

ABSTRACT: Static indentation and impact tests have been performed on thin plain and stiffened CFRP panels. The impact perforation threshold level may be estimated from the work done by the indentor during a static indentation test. Normalised damage maps produced from impact test data suggest that both plain and stiffened panels have similar damage profiles. A means of predicting the behaviour of large panels based on laboratory test data is suggested.

1 INTRODUCTION

Aerospace structures are subjected to impact during service and to mishandling during manufacture and maintenance. The damage induced provides a complex interaction of a range of failure mechanisms. This often necessitates further testing and evaluation for changes of material, lay-up or geometry. We have recently shown that some of these issues may be resolved by producing damage maps based on the impact performance threshold energy for plain CFRP panels (Found et al. 1997).

Whilst the low-velocity impact behaviour of plain laminates has been investigated for over two decades only limited studies have been undertaken on stiffened sections or structures. These show that significant improvements in resistance to impact damage may be achieved with the use of a thin membrane to absorb the energy with the structural stiffness being provided by careful design and positioning of local stiffeners (Maddam and Sutton 1988, Cheung and Scott 1995, Greenhalgh et al. 1996). The presence of localised stiffeners increases the available elastic energy which can be absorbed before fracture commences (Davies and Zhang 1995). They also produced force maps, which they show to be independent of plate size effects, relating damage area to the maximum force. However our work indicates that energy maps may be more appropriate since it is possible to identify and predict damage mechanisms over a wider range of impacts based on the static perforation energy (Found et al. 1997).

The increasing use of composite materials in aerospace structures is producing a demand for more competitive designs and a reduction in manufacturing costs. In order to cut product lead-times and reduce manufacturing costs it is necessary to be able to predict the behaviour of large structures from small-scale laboratory tests. This paper suggests a means of predicting the impact perforation threshold energy, and hence to identify the principal failure mechanisms for both plain and stiffened panels, based on the static perforation energy associated with a static indentation test. Analysis of data from indentation and dropweight impact tests offers a possible solution for predicting the behaviour of large plain and stiffened CFRP panels subjected to impact.

2 EXPERIMENTAL

CFRP panels nominally 350 x 350 mm, stiffened with three parallel T-blades of 100 mm pitch were supplied by Hurel-Dubois UK. The blades measured 25 mm wide x 12.5 mm deep and the webs were produced from two plies back-to-back with the ends bent at 90° to form a single thickness for the flange. Plain panels of the same three-ply lay-up were also supplied for comparison. The material was a five-harness satin weave carbon fibre preimpregnated with an epoxy resin designated 914C-713-40 and supplied by Hexcel Composites. The panels were laid up in three plies as (0/90, ±45, 0/90) and 0.90 stiffeners added, each panel being moulded in one shot by Hurel-Dubois at a nominal 58% fibre volume fraction.

An instrumented dropweight impact rig was used for both static indentation tests and low velocity impact tests. The panels were clamped to the same ring pressure using two annular rings ranging from 100 to 300 mm internal diameter, the lower ring having slots to accept the webs of the stiffeners. Dropweight impact tests were conducted from a height of 0.5 m to produce an impact velocity of the impactor of about 3 m/s. The impact forces and displacements were obtained from data that was processed through a digital low-pass filter set at a cut-off frequency of 3.5 kHz. (Found et al. 1998b). Static indentation and impact tests were

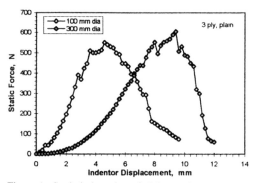

Figure 1. Static indentation of plain panels.

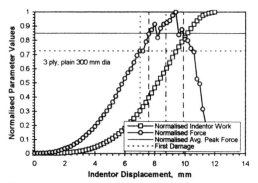

Figure 2. Normalised static indentation plot for plain panel.

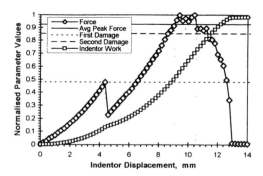

Figure 3. Normalised static indentation plot for stiffened panel in-line with stiffener.

performed on the stiffened panels at three different locations namely; in the bay between the stiffeners, at the toe of stiffeners and directly in-line with stiffeners. Damage was assessed using x-radiography and microscopy techniques in order to determine the principal failure mechanisms of backface cracking, delamination and permanent indentation of the frontface.

3 RESULTS AND DISCUSSION

Figure 1 shows typical force displacement plots for three-ply plain panels of 100 and 300 mm diameter obtained from static indentation tests. There are a number of similarities between the plots in terms of initial damage and peak force with the larger panel having slightly higher values. However in both cases the peak force was sustained over an indentor displacement of approximately 2 mm, which has also been observed for thicker panels (Found et al. 1997). The main differences are reflected in the initial slopes of the profiles where the larger more flexible panel is subjected to membrane bending effects. Also beyond the peak force there is a gradual reduction in force with increasing displacement for the 100 mm panel, whilst for the 300 mm panel there is a rapid reduction in force due to greater stored energy, leading to perforation of the panels.

In an attempt to identify possible relationships in terms of perforation of the panel the data for the 300 mm plain panel shown in Figure 1 has been normalised to the peak static force as presented in Figure 2. Also shown in Figure 2 is the normalised work done by the indentor during the test. Often the peak force is not as clearly defined as indicated here and therefore an average peak force is determined. This is obtained from the area under the force-displacement curve from initial damage load to the corresponding position on the downside of the plot. The initial damage load is identified by the first change in the upside of the plot which occurs at approximately 72% of the peak force shown in Figure 2. The static perforation energy of the panels may be estimated in terms of maximum work done by the indentor to produce upperbound and lowerbound values. The former is obtained from the crossover of the two plots in Figure 2 which occurs at approximately 82% of the maximum work done to give an estimated static perforation energy of 2.50 J. The lower value is determined from the work done by the indentor at the median displacement associated with the average peak force which occurs at approximately 60% of the maximum work done to produce an estimated static perforation energy of 1.82 J. The initial damage is in the form of backface cracking and remains dominant up to about the average peak force when delaminations become dominant. At the start of the downside of the load-displacement plot fibre fracture is significant leading to perforation of the panel.

Figure 3 shows a plot of normalised peak force for indentation in-line with the stiffener of a 300 mm diameter stiffened panel. Here the first damage load is associated with vertical cracking of the web of the stiffener and occurs at approximately 50% of the peak force. The second damage load is related to the initial damage in the panel skin (as also shown in Figure 2). The crossover of the two plots in Figure 3 also occurs at approximately 82% of the maximum work done by the indentor but gives a much increased upperbound estimate of the static perforation energy of 5.58 J when the indentor is in-line with the stiffener. For the 300 mm

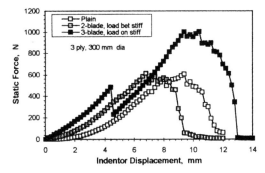

Figure 4. Static indentation of plain and stiffened panels.

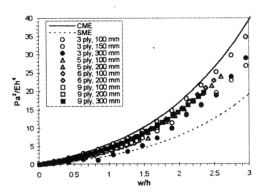

Figure 5. Non-dimensional force-displacement analysis of plain panels.

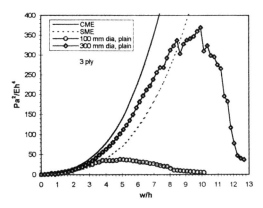

Figure 6. Force-displacement plots for plain panels compared with plate theory.

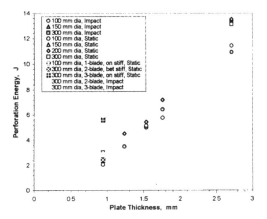

Figure 7. Static and impact perforation energies for plain and stiffened panels.

stiffened panel it is also possible to test in the bay of the panel, i.e. between two stiffeners and the results are presented in Figure 4 together with the data from Figures 2 and 3. Whilst Figure 4 shows that the peak force is similar for the plain panel and testing between the stiffeners of a stiffened panel a greater indentor displacement is required to perforate the plain panels. Hence the static perforation energy is slightly higher for the plain panels (2.50 J), due to their ability to store more energy, compared with the stiffened panels (2.44 J).

In order to identify the effect of panel thickness and diameter on the force-displacement behaviour for plain panels we have used an approximate analysis (Timoshenko 1970) for thin isotropic panels which accounts for membrane bending and is given by:

$$\frac{w}{h} + A\left(\frac{w}{h}\right)^3 = B.\frac{Pa^2}{Eh^4} \quad (1)$$

where P is the force, E is Young's modulus, a is the panel radius, h is the thickness, w is the central displacement and A and B are constants. Encouragingly the data for all the tests almost collapse onto a single curve as shown in Figure 5. Furthermore, the data are bounded

by two of the four end conditions considered by Timoshenko (1970), namely an upperbound when the panels are clamped with moveable edge (CME) and a lowerbound of simply supported with moveable edge (SME). Clearly this first estimate will permit the prediction of the force-displacement response of large structural panels. Further investigation shows that the force plots only deviate from the isotropic plate theory after initiation of damage as identified by variation of the plots as shown in Figure 6. It is thought that by modification of the plate theory to incorporate the effect of stiffeners that a similar response to Figure 5 may be achieved with stiffened panels.

The static perforation energy also appears to be related to the impact perforation threshold energy for plain panels as shown in Figure 7. There is some scatter in the data mainly arising from the variation in the static perforation energy which appears to have a range of values as discussed above. In this plot the upperbound value has been used to estimate the static perforation energy. Figure 7 also identifies the influence of panel diameter, with importantly the larger 300 mm panels having greater impact perforation energies than the smaller 100 mm panels. This highlights the very limited use of such small samples for trying to predict the response of large structural panels. It also shows that large structural panels should be capable of achieving greater damage tolerance than is often realised. To data we have only tested stiffened panels of one thickness, i.e. of three-ply material and these results are included for comparison purposes. We would expect similar increases in perforation energy for tests on thicker panels when the indentor is in-line with the stiffeners.

We have previously presented (Found et al. 1997) energy maps as a means of identifying the different mechanisms that develop from initial damage to perforation for plain panels subjected to low velocity impact. In this paper Figures 8 and 9 represent energy maps relating to the projected delamination area and total backface crack length (i.e. the combined major cracks in the 0° and 90° directions) respectively for both plain and stiffened panels of 300 mm diameter. Figure 8 shows that whilst the maximum delamination area is not significantly different for each of the test conditions, that when impacted in-line with a stiffener the largest damage occurs at a much higher energy and that the perforation energy is also far greater. Similar trends are also observed in Figure 9 for backface cracking where the damage relates to that in the panel skin. For a stiffened panel subjected to impact in-line with a stiffener, there is additionally a vertical crack in the web of the stiffener which occurs prior to the damage in the panel skin as earlier identified for static indentation tests. Figures 8 and 9 confirm that stiffened panels subjected to impact between the stiffeners suffer similar levels of damage to that of plain panels.

The energy maps shown in Figures 8 and 9 may be analysed differently by normalising both the damage parameters and the incident kinetic energy associated with the onset of perforation of the panels as presented in Figures 10 and 11. We determine the onset of perforation from examination of x-radiographs of the projected delamination area and it usually coincides with daylight just being visible through the damage site. This appears to be justified since increasing levels of perforation do not produce further increase in size of the main damage parameters. It is interesting to note that in both Figures 10 and 11 that all the data approximately collapses onto a common profile and hence the stiffened panels are subjected to proportionally similar levels of damage from onset of damage to perforation as the plain panels. By careful examination of these figures there appear to be similar changes in mechanisms for the stiffened panels at about 50 and 70% of the perforation energy as previously observed for plain panels (Found et al. 1998a). Thus the damage maps produce four different regions which can be predicted from knowledge of the impact perforation energy. These are (1) from onset of damage up to 50% of the perforation threshold energy (PTE), (2) from 50 to 75% of the PTE, (3) from 75% to the onset of perforation and (4) from onset to complete perforation of the panel.

4 CONCLUSIONS

Static indentation and dropwright impact tests have been conducted on thin plain and T-stiffened CFRP panels. The static perforation energy may be estimated from the work done by the indentor and the values used to estimate the impact perforation energy. The response of stiffened panels loaded between the stiffeners is similar to that of plain panels. As expected, the perforation energy is significantly increased when tests are performed in-line with the stiffeners. Similarly the maximum damage in the form of delamination and backface cracking occurs at higher energies than for other panel test conditions. When the impact damage parameters are normalised with respect to the perforation threshold energy the data for tests on both plain and stiffened panels fall on similar curves. This suggests that similar procedures may permit the prediction of damage in large thin CFRP panels based on knowledge of the impact perforation energy. Also means of predicting the static response of plain panels has been suggested based on isotropic plate theory. Furthermore the impact perforation energy may be estimated from static indentation tests. Thus by combining some of the procedures outlined in this paper goes some way towards being able to predict the response of large, thin plain or stiffened CFRP panels based on laboratory-scale tests.

REFERENCES

Cheung A.K.H. & Scott M.L. 1995. Low velocity impact and static indentation of thin, stiffened, composite panels. 2nd Pacific Int. Conf. *Aerospace and Tech.*, 631-638.

Davies G.A.O. & Zhang X. 1995. Impact damage prediction in carbon composite structures. *Int. J. Impact Engg.* 16(1), 149-170.

Found M.S., Howard I.C. & Paran A.P. 1997. Size effects in thin CFRP panels subjected to impact. *Comp. Struct.* 38, 599-607.

Found M.S., Howard I.C. & Paran A.P. 1998. Impact perforation of thin CFRP Iaminates. Proc. *12th Euro Conf. on Fracture* (in press).

Found M.S., Howard I.C. & Paran A.P. 1998. Interpretation of signals from dropweight impact tests. *Comp. Struct.* (in press).

Greenhalgh E.S., Bishop S.M., Bray D., Hughes D., Lahiff S. & Millsor B. 1996. Characterisation of impact damage in skin-stringer composite structures. *Comp. Struct.* 36, 187-207.

Maddam R.C. & Sutton J.O. 1988. Design, testing and damage tolerance study of bonded stiffened wing covers. 29th Struct. Dynam. and *Maths Conf.* AIAA/ASME/ASCE/AHS, 623-630.

Timoshenko S.P. & Woinowsky-Krieger S 1970 2nd Edn McGraw-Hill, 396-415.

Experimental Techniques and Design In Composite Materials 4, Found (Ed.)
© *2002 Swets & Zeitlinger, Lisse, ISBN 90 5809 370 0*

Simulation of the impact response of composite aircraft substructures

Th. Kermanidis, G. Labeas & S. Pantelakis
Laboratory of Technology and Strength of Materials, University of Patras, Greece

D. Kohlgrueber
Deutsches Zentrum fuer Luft- und Raumfahrt - DLR, Stuttgart, Germany

& J. Wiggenraad
National Aerospace Laboratory - NLR, Amsterdam, The Netherlands

ABSTRACT: In terms of crashworthiness, advanced composite structures require special design compared to metal structures, in order to overcome the drawback of their high degree of brittleness and enable superior impact behaviour. The 'tensor skin' concept, which was developed by NLR and originally suggested as energy absorbing composite system for the sub floor structure of helicopters subjected to water impact, is studied in the present paper. In order to identify the ability of the 'tensor skin' concept to transfer impact loads and absorb impact energy, static crush and dynamic drop tests were performed on the practical realization of the concept, which is a corrugated sandwich configuration. The modelling methodology and the numerical simulations of selected proof tests of the 'tensor skin' system, using the Finite Element code PAM- CRASH, are presented. A good correlation between experimental and numerical results was achieved.

1 INTRODUCTION

The utilisation of advanced composite materials on airframe structures has been increasingly accepted during the past decades, due to their high stiffness to weight and strength to weight ratios. When composite materials are used in fuselage subfloor components, the proof of their ability to withstand impact loading, which occurs in crash situations, is a major design goal, provided the limited ability of composite systems in absorbing impact energy. Within the European research programme 'Design for Crash Survivability - CRASURV', which is a running RTD project partially funded by the European Union in the Aeronautics Area of the Industrial and Materials Technologies (BRITE/ EURAM) programme, various crash-worthiness composite components, i.e. beams, panels and cruciforms are developed. These components are to be implemented in subfloor boxes of helicopter and aircraft fuselages, as energy absorbing elements. In order to verify their ability in energy absorption, the developed components are experimentally tested. However, tooling, manufacturing and testing of components is quite expensive. To allow parameter studies and minimise the tests, prediction techniques for crash behaviour of composite structures are required.

The widely applied Finite Element (FE) method is a suitable tool for the modelling and the simulation of the initiation and propagation of the different types of failure within an impacted structure. However, less reliability exists in the utilisation of such codes in the case of composite structures, compared to the metallic ones. Major issues of the numerical simulation of composite

structures under impact are the proper representation of the composite material behaviour, the prediction of the complex failure modes and the development of adequate FE meshes which will lead in accurate results in reasonable solution times.

In the present work the non-linear dynamic FE code, PAM-CRASH is used. The experimental tests which are simulated, are the static crush tests of the 'tensor skin' strip and panel and the dynamic drop tests of the 'tensor skin' panels performed by NLR [3] and DLR [8]. The simulation methodology is described in detail. The results of the simulations are generally in good agreement to the experimental ones. It arises that the investigated structures can transfer the impact loads properly.

2 NUMERICAL SIMULATION METHOD

The subcomponents studied in the present work are two realizations of the 'tensor skin' concept, having the sandwich configuration shown in figure 1. The sandwich consists of a corrugated skin surrounded by an inner (hat shaped) and an outer face. The material systems utilised for the inner and outer facings are Carbon-Aramid/epoxy hybrid fabric (trade name Hexcel 73210-2-1220-F155-45%) and Aramid/epoxy fabric (Hexcel F-155-49-285-52%), respectively. The corrugated core of the panel is made of Dyneema fabric, which is made of a high performance polyethylene fibre, (DSM SK 60-132 TEX) with RTM resin (Ciba Geigy 5052). Experimental data from tension, compression and shear coupon tests for these materials are taken from [5] and [6].

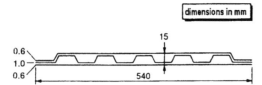

Figure 1. Cross section of a corrugated tensor skin panel.

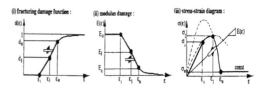

Figure 2. Fracturing damage function, modulus degradation and stress-strain diagram [4].

For the simulation of the crush tested composite substructures the PAM-CRASH FE code is used. One of the major subjects for a successful numerical simulation of an impacted structure, using non-linear dynamic FE codes, is the development and calibration of suitable material damage models. Such models have to represent properly the material stiffness and strength degradation.

The PAM-CRASH code enables the modelling of composite layered structures, using four node shell elements with one integration point per layer, combined to a material type, coded 'Material 130', which represents the anisotropic behaviour. Elastic fiber-matrix behaviour with damage can be modelled by 'Material 130'. Different material properties can be defined for each layer, requiring for each of them stiffness, strength, and damage progression data. For each layer the initial undamaged in-plane stiffness properties E_{11}, E_{22}, G_{12}, and ν_{12} should be provided, for the calculation of the initial modulus matrix $\mathbf{C_0}$. A damage function $\mathbf{d}$ [4], enables the representation of the degradation of the initial modulus matrix $\mathbf{C_0}$, when an initial undamaged phase is exceeded. The modulus matrix behaves according to the formulae:

$$\mathbf{C}(d) = \mathbf{C_0} \, (1\text{-}d) \qquad (1)$$

The damage function $\mathbf{d}$ is a scalar parameter that depends upon strain:

$$d(\varepsilon) = d_v(\varepsilon_v) + d_s(\varepsilon_s) \qquad (2)$$

where d_v is the volumetric damage due to a volumetric equivalent strain ε_v and d_s is the shear damage due to a shear equivalent strain ε_s. The scalar parameter ε_v represents the first invariant of the volumetric strain tensor, while the scalar ε_s represents the second invariant of the deviatoric strain tensor. For the simplification of an uniaxial test, the equivalent volumetric and shear damage values are defined as:

$$\varepsilon_v = \varepsilon_{kk} = (1\text{-}\nu_{12}\text{-}\nu_{13})\varepsilon_{11} \qquad (3)$$

$$\varepsilon_s = [(1/2)\,e_{ij}\,e_{ij}]^{1/2} = (\varepsilon_{11}/\sqrt{3})\,(1 + \nu_{12} + \nu_{13} + \nu_{12}\,\nu_{13} + \nu_{12}^2 + \nu_{13}^2)^{1/2} \qquad (4)$$

In equations 3 and 4, ε_{kk} is the trace of the total strain tensor and e_{ij} are the components of the deviatoric strain tensor. The implemented damage law in PAM-CRASH assumes that the fracturing damage parameter $\mathbf{d}$ is zero

for an equivalent strain between zero and ε_i (figure 2-i). After the value ε_i is reached, the fracturing damage factor $\mathbf{d}$ grows linearly between the values ε_i and ε_1. Between ε_1 and ε_u the damage factor $\mathbf{d}$ grows linearly again, with a different slope. The damage parameters which correspond to the strains ε_i, ε_1 and ε_u are d_i (which always has a zero value), d_1 and d_u respectively, where d_u is the stage of the ultimate damage. The elasticity modulus is assumed to degrade according to fig. 2-ii and is related to uniaxial data according to fig. 2-iii.

The calibration of the material damage models can be performed either using the tension-compression stress-strain curves to introduce only volumetric damage, or using the shear stress-strain curves to introduce only shear damage. In the former case, the slopes can be calculated from the tension-compression uniaxial data as:

$$E_{v0} = \frac{\sigma_i}{\varepsilon_{vi}} \qquad E_{v1} = \frac{\sigma_1}{\varepsilon_{v1}} \qquad E_{vu} = \frac{\sigma_u}{\varepsilon_{vu}}, \qquad (5)$$

The volumetric damage values then arise:

$$d_{v1} = 1 - \frac{E_{v1}}{E_{v0}} \qquad d_{vu} = 1 - \frac{E_{vu}}{E_{v0}} \qquad (6)$$

The use of either only volumetric, or only shear damage, can lead to the successful representation of the experimental tension and compression data. However, this is not the case for the shear behaviour, which is overestimated or underestimated, if only volumetric or only shear damage is used. To face this problem, the damage models are calibrated starting with the calculation of the shear damage from the shear coupons tests. After selecting a set of values for the strains ε_i, ε_1 and ε_u the slopes G_i, G_1 and G_u and the corresponding shear damage parameters d_1 and d_u are calculated. Different combinations of ε_i, ε_1 and ε_u were tried to achieve a good representation of the τ-γ curves for each material. The obtained results, using shear damage only, have shown an enormous overestimation of the tension/compression strengths, because of the fact that the composite fabrics which are used here, have a completely different behavior in tension/compression (ε_{max} is between 1% and 3%) and shear (γ_{max} is between 10% and 18%). For this reason, volumetric damage parameters are introduced afterwards, to cut-off the final strength and match the tension/compression data.

An additional parameter which should be introduced into the material model is the element elimination strain.

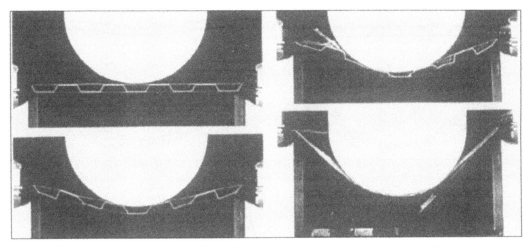

Figure 3. Deformed shapes of the 'tensor skin' strip static crush test, at various time intervals [3].

When the equivalent strain value at an element reaches the defined elimination strain, the code automatically eliminates the corresponding element.

The bounded interfaces in the corrugated tensor skin (figure 1) are modelled using a tied contact algorithm, coded in PAM-CRASH as contact 'type 2'. This algorithm requires the input of normal N_s and shear T_s strengths of the bond and enables the contact failure, after the contact force of the tied nodes is exceeded. The failure occurs [4] when:

$$\left(\frac{N}{Ns}\right)^2 + \left(\frac{T}{Ts}\right)^2 \geq 1 \tag{7}$$

For the modelling of the contact between the structure interfaces, the self contact algorithm, coded in PAM-CRASH as contact 'type 26' is applied. This contact algorithm automatically searches for elements contacting each other and introduces an internal contact force between these elements.

The tools which introduce the deformation are modelled as moving rigid walls with infinite or finite mass. The real tool velocity in the static crush tests is constant with values between 5mm/min and 20mm/min. As velocities of this magnitude would lead to very long simulation times, the rigid walls velocity is considered higher. However, in order to avoid a sudden impact at the beginning of the simulation, a linearly increasing rigid wall velocity for all the simulations of static tests is ranging from zero (at time=0) up to 1m/min (at time=1sec).

3 SIMULATION OF THE TENSOR SKIN STRIP STATIC CRUSH TEST

'Tensor skin' strips of small width (135mm) and 540mm length, having the cross-section of figure 1, were manufactured. The outer and inner skins, which have thick-

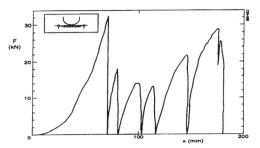

Figure 4. Measured tool force vs. time of the 'tensor skin' strip static crush test [3].

ness 0.78mm and 0.88mm, respectively, were made in autoclave process at a [±45/0/±45] lamination. The middle Dyneema facing, was manufactured using R TM at a thickness of 1.11mm and [±45] lamination. The fibre material of the Dyneema layers is polyethylene which allows much higher strains to failure than the conventional fibre systems. As a result, upon loading, they unfold and deflect by forming "plastic hinges" before they get stretched and fail in tension, allowing the absorption of the impact energy. To identify the load carrying capability of the system, a static crush test was performed, as shown in figure 3. The obtained force-displacement curve is shown in figure 4. The failure is initiated from the outer and inner facings, at a force of 31kN (fig. 4) and consequently, unfolding and progressive failing of the Dyneema layers takes place resulting in local peaks of the transverse force.

For the simulation of this test, a relatively detailed FE mesh was developed, comprising 2873 nodes and 2616 shell elements. The load introduction device has been modelled by a rigid wall of infinite mass, moving with a linearly increasing velocity. Both the normal N_s and

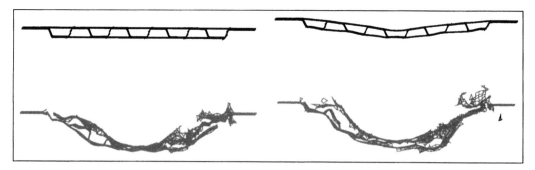

Figure 5. Calculated deformed shapes of the 'tensor skin' strip static crush test, at various time intervals.

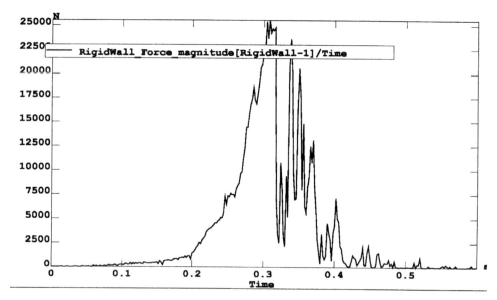

Figure 6. Rigid wall force vs. time of the 'tensor skin' strip static crush test.

shear T_s strengths of the tied interfaces were assumed to be 5kN. A self contact algorithm was applied for all the elements of the model. In figure 5 the calculated deformed shapes of the 'tensor skin' strip static crush test, at various time intervals from the initial unde-formed state, up to the state of the final failure, are pre-sented. In figure 6, the calculated rigid wall force versus time of the 'tensor skin' strips is shown.

Comparing the experimentally obtained deformed shapes (fig. 3), to the calculated deformed shapes, (fig. 5), it arises that the simulation predicts well the failure process. However, the final phase of the test is the com-plete unfolding of the Dyneema layers, which is not properly predicted by the simulation (fig. 5). The reason is that the accumulated damage at the left edge of the outer facing reaches its ultimate value, causing the total structure failure, before the complete unfold of the mid-dle face.

Figure 7. The 3-D 'tensor skin panel'.

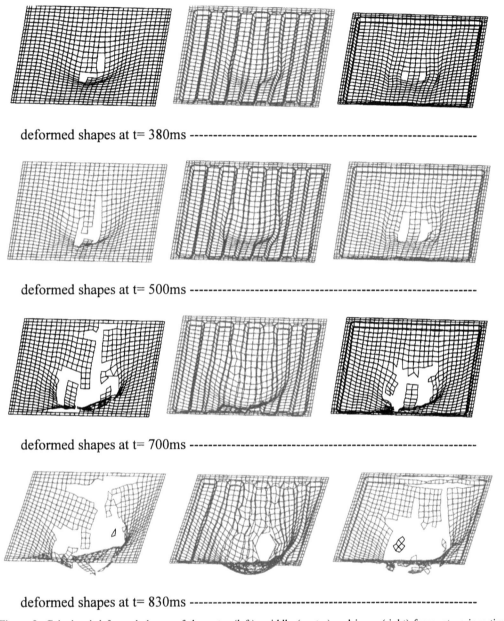

deformed shapes at t= 380ms ---

deformed shapes at t= 500ms ---

deformed shapes at t= 700ms ---

deformed shapes at t= 830ms ---

Figure 8. Calculated deformed shapes of the outer (left), middle (center) and inner (right) faces, at various time intervals, of the 'Mid45St' static crush test.

From figures 4 and the corresponding displacement versus time curve, which is not shown here, it arises that the outer and inner faces fail at 0.4s, which corresponds at a the rigid wall force of 25.2kN and rigid wall displacement of 62mm. The time for the final panel failure is 0.49s, which corresponds to a rigid wall displacement of 124mm. From figure 4 the measured tool force at the time of the outer and inner faces failure initiation is 31kN at 70mm tool transverse displacement, values which are quite close to the prediction. However, in the simulation, the Dyneema layers fail before they have totally unfolded, as mentioned before, at a maximum tool displacement of 124mm, which is relatively lower than the experimentally measured tool displacement of about 180mm.

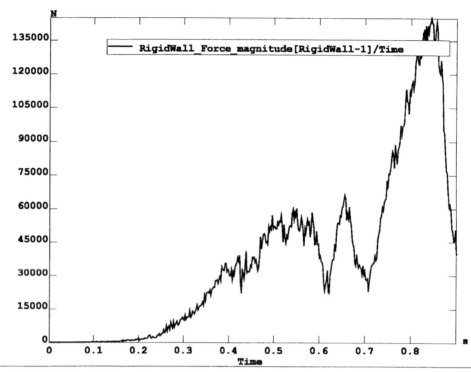

Figure 9. Rigid wall force vs. time of the Mid45St static crush test [3].

4 SIMULATION OF THE 'TENSOR SKIN' PANEL STATIC CRUSH TEST

The three-dimensional 'tensor skin' panel is a 540x540mm square panel, with the cross section shown in figure 1. A photo of the panel is shown in figure 7. The outer face is made from three Aramid/epoxy layers at [±45/0/±45] lay-up and the inner hat shaped face is made of three hybrid Carbon-Aramid/epoxy layers, at the same lay-up. The corrugated core consists of three Dyneema layers at [±45]₃ lay-up, manufactured by the RTM technique, which is especially suitable for the fabrication of corrugated cores with many folds. The panel is clamped at all four edges, and imposed to transverse loading by a spherical tool of 150mm diameter. The test is coded 'Mid45St'. For the simulation of this experiment a model of the panel was developed. All the geometrical data for the modelling have been taken from [7].

The FE mesh is relatively rough comprising 3300 elements (3837 nodes). Both the normal N_s and shear T_s strengths of the tied interfaces were assumed to be 5kN. The element elimination strains were assumed to be 0.6 for Dyneema and 0.2 for the Aramid and Carbon/Aramid fabric material systems.

The calculated deformed shapes of the outer, middle and inner faces, of the 'Mid45St' static crash test at various time intervals are shown in figure 8. The calculated rigid wall force versus time are shown in figure 9. During the test the two faces failed early, as expected, and the corrugated core unfolded and stretched to fail only at very high load.

In figure 10 the experimentally measured tool force is plotted versus tool displacement. From figures 9 it arises the outer and inner faces fail at a rigid wall force of 36kN (at 0.39s) and a rigid wall displacement of 39mm. The time for the final Dyneema failure is 0.84s, which corresponds at a the rigid wall force of 149kN and rigid wall displacement of 178mm. From figure 10 it can be concluded that the outer and inner faces fail at a tool force of 31.8kN and rigid wall displacement of 42mm, while the Dyneema layer fails at a tool force of 171kN and rigid wall displacement of 149mm. Comparing the experimental and calculated values, a very good agreement is observed, especially concerning the peak loads. Finally, in figure 11, the final deformed shapes of the outer and inner faces of the panel are shown. A comparison between figures 10 (final row) and 11 shows that the failure modes of the three panel layers are simulated successfully.

5 SIMULATION OF THE MASS DROP TESTS OF THE 'TENSOR SKIN' PANEL

The square 'tensor skin' panels described in the previous paragraph, were also dynamically tested. The panels

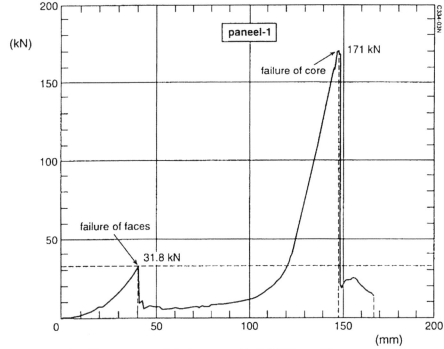

Figure 10. Measured tool force versus tool displacement of the Mid45St test [3].

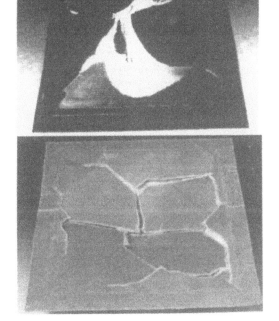

Figure 11. Failure of the inner (up) and outer (down) faces from the 'Mid45St' test.

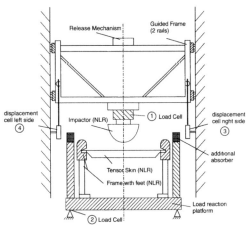

Figure 12. Experimental set-up for the dynamic mass drop tests [8].

were clamped at all edges and an impactor of a specified mass dropped on the panels centrally, at a pre-defined initial velocity. The two dynamic tests are coded 'TSDy2D' and 'TSDy3D'. All the experimental data which are required for the simulations were taken from [8]. The 'TSDy2D' panel has two corrugated Dyneema layers of a total thickness of 0.74mm and is impacted by a 59.55kg mass at an initial velocity of 7.244m/s. The

'TSDy3D' panel has three corrugated Dyneema layers of a total thickness of 1.11mm and is impacted by a 78.95kg mass at a velocity of 7.534m/s. The experimental set-up is shown in figure 12.

For the simulation of the two dynamically tested corrugated panels, the same FE mesh to that developed for the simulation of the static crush test is used. The tools were simulated as spherical rigid walls with their corresponding masses and initial velocities. The connection between the three layers of the panel are considered as tied interface with failure. The normal N_s and the shear T_s strengths of the tied interface are assumed to be 3kN,

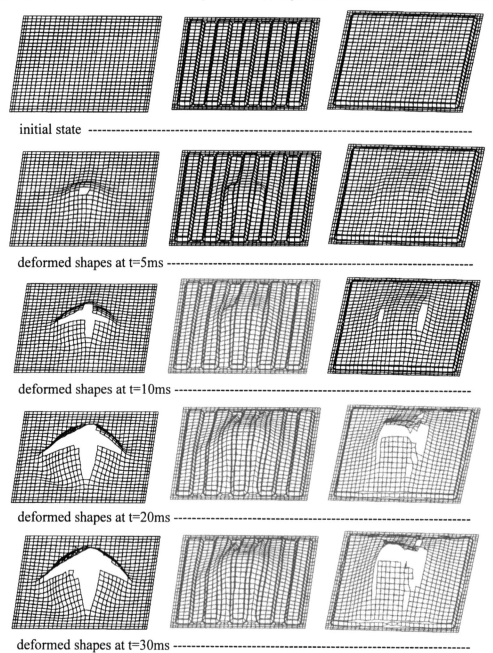

initial state ---

deformed shapes at t=5ms ---

deformed shapes at t=10ms --

deformed shapes at t=20ms --

deformed shapes at t=30ms --

Figure 13a. Deformed shapes of the outer (left), middle (center) and inner (right) faces, at various time intervals, of the 'TSDy3D' dynamic test.

124

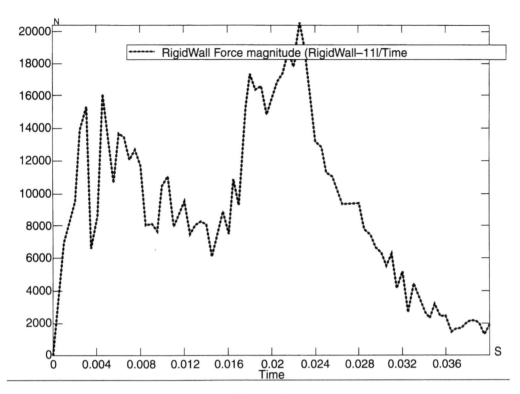

Figure 13b. Rigid wall force vs. time of the 'TSDy3D' dynamic test.

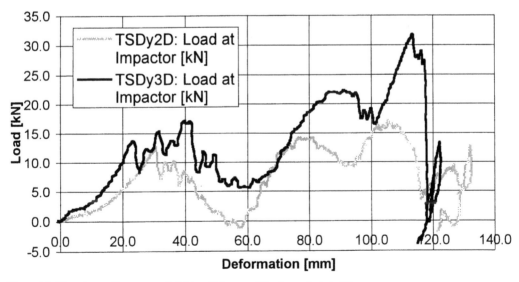

Figure 14. Experimentally measured rigid wall forces of the dynamic tests [8].

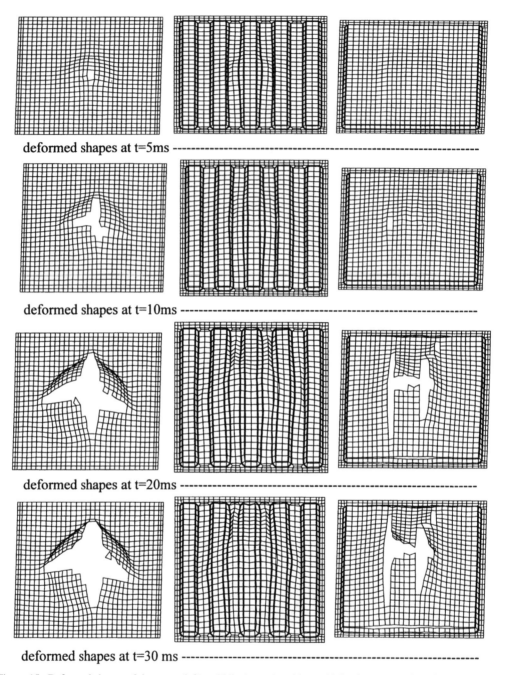

deformed shapes at t=5ms --

deformed shapes at t=10ms --

deformed shapes at t=20ms --

deformed shapes at t=30 ms ---

Figure 15. Deformed shapes of the outer (left), middle (center) and inner (right) faces, at various time intervals, of the 'TSDy2D' dynamic test.

which is quite low compared to the values used in the simulation of the static test, as it was observed that in the dynamic tests the interlayer bonding fails almost completely and quite early. The element elimination strains were assumed to be 0.6 for Dyneema and 0.1 for the Aramid and Carbon/Aramid fabric material systems. Self contact (type 26) was considered for the whole structure.

126

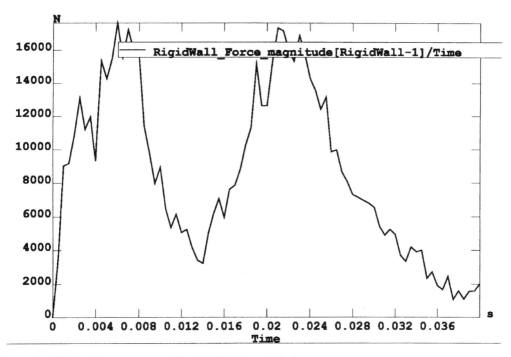

Figure 16. Rigid wall force vs. time of the 'TSDy2D' dynamic test.

Figure 17. Failure mode of the outer face of the 'TSDy2D' panel.

The results of the simulations which are described in the following, include calculated deformed shapes at various time intervals, calculated rigid wall (rw) forces versus time and some of the corresponding experimental results, from [8].

The deformed shapes of the three panel faces, of the 'TSDy3D' dynamic test, at various time intervals are shown in figure 13a and the rigid wall force versus time curve in figure 13b. The experimentally measured tool force versus tool displacement for both the 'TSDy2D' and 'TSDy3D' tests are shown in figure 14. From a comparison between experimental and calculated results of the 'TSDy3D' test the following comments can arise: a) the failure modes of the three layers of the 'TSDy3D' test are well simulated; b) in the simulation the maximum rw displacement is 96mm and occurs at 25ms, while the measured values are 122.6mm and 30ms, respectively; c) in the simulation the first rw peak force is 15.3kN, at 5.5ms, at 37mm displacement, while the measured values are about 17.5kN, at 40mm displacement, respectively; d) in the simulation the second rw peak force is 17.5kN, at 18ms, at 82mm displacement, while the measured values are about 22.5kN, at 88mm displacement, respectively; e) in the simulation the third rw peak force is 20.5kN, at 23ms, at 92mm displacement, while the measured values are about 31.7kN, at 115mm displacement, respectively.

In figure 15, the deformed shapes of the three panel faces, of the 'TSDy2D' dynamic test, at various time intervals are shown. The rigid wall force versus time is shown in figure 16, while a photo of the failed panel after dynamically tested appears in figure 17.

Comparing the experimental to the calculated results, the following comments can arise: a) the failure modes of the inner and outer layers of the 'TSDy2D' test are generally well simulated; b) the Dyneema layer fails partially in the test which is not predicted in the simulation; c) in the simulation the maximum rw displacement is 100mm and occurs at 26ms, while the measured values are 132.3mm and 32ms, respectively; d) in the simulation the first rw peak force is 13kN, at 2.5ms, at 25mm displacement, while the measured values are about 13kN, at 30mm displacement, respectively; e) in the simulation the second rw peak force is 16.8kN, at 6ms, at 50mm displacement, while the measured values are about 14.5kN, at 78mm displacement, respectively; f) in the simulation the third rw peak force is 16.5kN, at 22ms, at 98mm displacement, while the measured values are about 17kN, at 108mm displacement, respectively.

6 CONCLUSIONS

The PAM-CRASH FE-code has been successfully applied for the simulation of the proof process of various realizations of the 'tensor skin' concept.

The crashworthiness capability of the various configurations, which were tested and simulated, was proved.

It is difficult to develop suitable composite material damage models to represent the material properties degradation with accuracy. However, the developed models were capable to represent successfully the material properties degradation.

The failure process of all the simulated structures was generally successfully predicted. A good agreement is observed between the calculated and measured tool forces.

Although good simulation results have been achieved in the present work, the simulation methodology, at present, should be considered only as a supporting tool for the design and proof process of composite crashworthiness structures under impact.

REFERENCES

Brite/Aero project BE95 2033, 1996, *"Design for Crash Survivability"*.

Lestari W., 1994, *"Crashworthiness Study of a Generic Composite Helicopter Subfloor Structure"*, NLR-TR 93590 L.

Michielsen A.L.P.J. & Wiggenraad J.F.M., 1997, *"Review of crashworthiness research of composite structure"*, NLR-CR 97046.

PAM-CRASH *Users Manual*, 1997, Engineering Systems International.

Ubels L.C., 1997, *"Initial Material data, Crasurv Task 1.1"*, NLR-TR 97308L.

Li Q.M., Mines R. & Birch R.S., 1997, *"Initial Material data for Composite Materials"*, Univ. Liverpool, IRC/151/97.

Michielsen A.L.P.J., 1997, *"Specification of sub-components and box structures"*, NLR-CR 97315.

Kohlgrueber D & Weissinger H., 1997, *"D.4.1.3. Results of Dynamic Tests of Tensor Skin Panels"*, DLR-IB 435-97/31.

Thuis H., Vries H. & Wiggenraad J., 1995, *"Subfloor skin panels for improved Crashworthiness of Helicopters"*, AHS 51st Annual Forum, Fort Worth, Texas.

Experimental Techniques and Design In Composite Materials 4, Found (Ed.)
© 2002 Swets & Zeitlinger, Lisse, ISBN 90 5809 370 0

Time-frequency Analysis of Impacts on a CFRP Panel

W.J. Staszewski, M.A. Maseras-Gutierrez, M.S. Found & K. Worden
Department of Mechanical Engineering, University of Sheffield, United Kingdom

ABSTRACT: The paper addresses the problem of impact damage detection in composite materials. Impact data is analysed using the continuous wavelet transform to reveal its varying time-frequency nature. The data is taken from a simple impact experiment in which piezoceramic sensors are used to collect the impact strain response. Also, the acceleration of the impacting body is measured. The study shows that the impact response data is affected by damage to the structure. Also, the analysis of the impacting body acceleration reveals the time-frequency features which can be associated with different types of damage to the structure.

1 INTRODUCTION

The objective of many current research activities in the area of composite damage detection is to develop different technologies and signal processing techniques suitable for on-line health monitoring systems. A number of different advanced signal processing techniques have been used to select appropriate features for damage detection.

It appears in practice that impacts with ground support equipment are the major cause of in-service damage to composite structures. Therefore impact identification analysis has a direct relevance to the problem of damage detection in composite materials. Previous study in this area include: FE modelling analysis (Staszewski 1996), experimental studies in metallic (Jones et al. 1995, Boller 1996) and composite (Weems et al. 1991, Gunther et al. 1992, Schindler et al. 1995) plates, neural network (Weems et al. 1991), Gunther et al. 1992, Schindler et al. 1995) and sensor optimisation (Staszewski et al. 1998) studies.

Significant impacts on composites can produce matrix cracking, delaminations, fibre fracture and perforation; the type of damage depends on the energy level of impact. It is important to establish different features in the data which can indicate particular damage. Often the structural integrity of the system is not affected by the damage introduced. The question remains whether such damage affects the impact data response. These aspects are the main objectives of the current work presented in this paper.

2 EXPERIMENTAL STUDY

An instrumented dropweight impact rig, briefly described in (Kumar et al. 1995, Staszewski et al. 1998)

was used for conducting the impact tests under circularly clamped loading conditions. The impact rig is equipped with an accelerometer, a strain-gauged load cell, a displacement transducer and opto-electronic triggering and timing sensors. The instrumented indentor is released from a pre-determined height by an electromagnetic switch and the data acquisition system is triggered when an aluminium flag, attached to the indentor assembly, passes the first opto-interrupter.

The test panel measured 340 x 340 x 2.5 mm and was cut from a laminate consisting of a carbon fibre fabric and a toughened epoxy resin. The laminate comprised of eight plies of preimpregnated material to give a quasi-isotropic lay-up [0/90, ±45, 0/90, ±45]s which was autoclave moulded by Hurel-Dubois UK. Four piezoceramic transducer elements measuring 27 mm diameter were adhesively bonded to the backface of the panel, at locations identified in Figure 1, in order to monitor the impact strain response.

Some preliminary impact tests were undertaken in order to ascertain the signal response from the piezoceramic sensors and to ensure that there was minimum crosstalk between the channels of the data acquisition system. Given that the initial out-put from the sensors was greatly in excess of that of the other transducers a simple voltage divider was incorporated into each of the piezoceramic sensors.

A series of impact tests were conducted on panels using hydraulically clamped rings of 300 mm internal diameter. The tests ranged from impact energies of 0.3 J, which was well below that needed to produce damage to beyond perforation of the panels at a maximum level of 10 J. On a separate fully instrumented panel, impact tests were undertaken at an initial energy of 0.3 J followed by a single impact at 9.6 J to cause perforation of the panel. The lower energy tests at 0.3 J

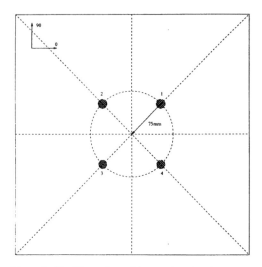

Figure 1. Backface of panel (sensor locations).

were again repeated on the same panel. All the unfiltered test data were analysed to see if the impact damage had affected the response of the piezoelectric sensor signals prior to the structural damage.

3 TIME-FREQUENCY ANALYSIS

Classical Fourier analysis provides a spectral representation of a signal which is independent of time and is therefore the analysis technique of choice for stationary time-series data. However, the data obtained from the experimental tests here exhibit clear non-stationary behaviour so Fourier analysis is inappropriate. A number of different time-variant techniques have been proposed for the analysis of non-stationary data; this paper uses wavelet analysis, which is a *time-scale* procedure.

The continuous time-scale wavelet transform is defined as,

$$W_{\psi}^{x}(a, b) = \frac{1}{\sqrt{a}} \int_{-\infty}^{+\infty} x(t) \psi^{*}\left(\frac{t-b}{a}\right) dt \quad (1)$$

where $\psi(t)$ is the analysing or *mother* wavelet used for the decomposition, b is a translation parameter indicating the locality in the time domain and a is a dilation or scale parameter which gives the frequency domain localisation. The normalisation by a factor $1/\sqrt{a}$ ensures that the integrated energy given by each wavelet $psi\_a, b(t)$ is independent of the dilation a. Equation (1) shows that, the continuous wavelet transform is a linear transformation that decomposes a given function $x(t)$ into a superposition of the elementary functions $\psi_{a,b}(t) = = \psi([t-b] a)$ derived from the analysing wavelet $\psi(t)$ by scaling and translation.

Any local features in the signal can be identified from the position and scale of the wavelets into which it is decomposed. In the frequency domain the wavelet

transform also allows the analysis to focus on a given bandwidth using a different size window depending. For the function $\psi(t)$ to qualify as an analysing wavelet, it must satisfy the appropriate admissibility condition

$$0 < C_g = \int_{-\infty}^{+\infty} \frac{|\Psi(f)|^2}{|f|} df < \infty \quad (2)$$

where $\Psi(f)$ is the Fourier transform of the function $\psi(t)$. The admissible function is in fact a band-pass filter. Many practical applications impose some other conditions such as decay rates at infinity, regularity and vanishing moments on the wavelets.

The analysis performed in this paper uses the Morlet wavelet defined by,

$$\psi(t) = e^{i2\pi f_o|t|} e^{-\frac{|t|^2}{2}} \quad (3)$$

More details about the wavelet transform, its properties and computation procedure can be found in Chui 1991, Staszewski et al. 1997. The wavelet transform is a two dimensional complex function which can be presented graphically in different ways. The application in this paper involves the representation of the transform in terms of its modulus. The phase of the transform is not used here.

4 RESULTS

The first part of the wavelet analysis considered the acceleration data from the impacting body for different levels of impact energy.

Figure 2 shows the wavelet analysis results for the energy level equal to 0.3 J. Here, the time-scale contour plot is given together with the analysed signal and its Fourier spectrum. Note that the logarithmic frequency axis in the spectrum is used for compatibility with the logarithmic scale axis of the wavelet filter. The contour

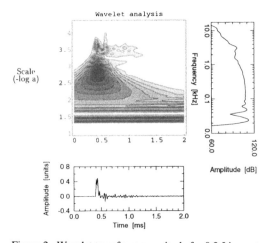

Figure 2. Wavelet transform magnitude for 0.3 J impact event.

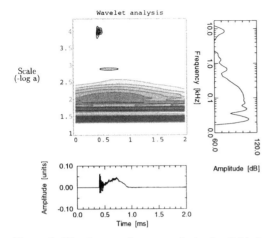

Figure 3. Wavelet transform magnitude for 2.44 J impact event.

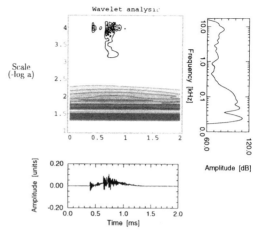

Figure 5. Wavelet transform magnitude for 7.83 J impact event.

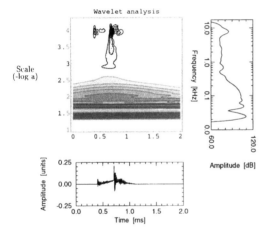

Figure 4. Wavelet transform magnitude for 4.24 J impact event.

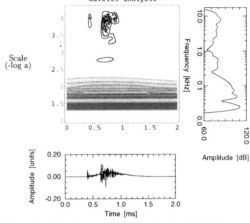

Figure 6. Wavelet transform magnitude for 9.6 J impact event.

plot of the wavelet transform magnitude displays a clear impact feature at 0.4 *ms* for frequencies in the range 0.7-0.8 *kHz*. This feature cannot be observed directly in the spectrum due to the non-stationary nature of the data. The bottom part of the contour plot represents the stationary part of the data within the frequency band of 8-80 *Hz*. The highest frequency components observed are at about 3-4 *kHz*. The impact energy used in this part of the experiment did not result in any damage to the composite panel.

A similar analysis for the impact energy level equal to 2.44 *J* can be seen in Figure 3. For the impact used in that experiment, matrix cracking has been observed (Maseras-Gutierrez). The major difference in the results, when compared with Figure 3, is the energy attenuation in the frequency bandwidth of 0.3-1.0 *kHz*.

Also, a new feature at 0.4 *ms* and 8-9 *kHz* can be observed.

The same featurs can be seen in the impact data for the energy level equal to: 4.24 *J*, 7.83 *J* and 9.6 *J* in Figures 4, 5 and 6 respectively. The 4.24 *J*, 7.83 *J* and 9.6 *J* impacts used in the experiment caused delamination, fibre fractures and full perforation, respectively. With the increase of the energy level, the amplitude of the wavelet transform exhibits clear high frequency (8-9 *kHz*) short-lived events in Figures 4-6.

The second part of the wavelet analysis involved the data taken from the piezoceramic sensors before and after the perforation damage produced by the 9.6 *J* Impact. (The 9.6 *J* impact produced a total backface crack length of 55 mm and a projected delamination area of 400 mm^2. The energy level of both impacts was

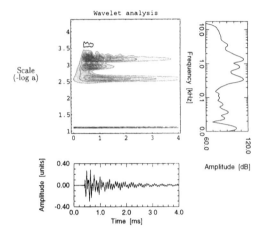

Figure 7. Wavelet transform from piezoelectric data before damage.

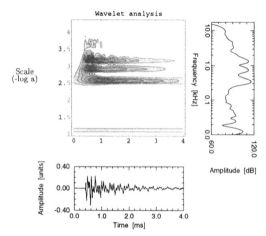

Figure 8. Wavelet transform magnitude from piezoelectric data after damage.

equal to 0.3 *J*.) The comparison between Figures 7 and 8 shows that the damage in the structure does affect the impact data taken by the piezoceramic sensors; clearly, the damping level after damage is smaller than before for one of the vibration modes of the panel.

5 CONCLUSIONS

The conclusions from this paper are clear. Because wavelet analysis is the appropriate tool for the analysis of non-stationary signals, and because impacts are manifestly non-stationary events, wavelet analysis is the correct tool for visualising the time-scale evolution of the impact signal. In the first sequence of signals analysed here, the wavelet magnitude shows the clear existence of features associated, not only with the impact, but with the consequences of that impact. There is a market progression in the wavelet plots as the impact energies increase and the consequent damages become more severe. In the second analysis, the wavelet shows clearly that damage modifies the behaviour of the sample under subsequent impacts. There are implications here for composite structural health monitoring. Firstly, the wavelet magnitudes are an informative feature, which could be used with an automatic pattern recognition system to associate damage states with impact time signals. Secondly, any such monitoring system must be able to take account of the modification to expected patterns if an early impact causes damage.

One caveat associated with the work here, is that the first sequence of time-histories were acceleration records from the drop weight. In practice, impacts will be from tool drop etc., and this information will not be available. However, it is anticipated that the same information will be available from strain time-histories measured on the structure, and the verification of this will form the next stage in this work.

REFERENCES

Boller Ch. 1996. Fundamentals of damage monitoring. *Smart Structures and Materials: Implications for Military Aircraft of New Generation. AGARD Lectures Series 205*, NATO.

Gunther M.F., Wang A., Fogg R.P., Starr S.E., Murphy K.A. & Claus R.O. 1992. Fibre optic impact detection and location system embedded in a composite material. *SPIE Proceedings of Fibre Optic Smart Structures and Skins IV*: 64-75.

Jones R.T., Sirkis J.S. & Friebele E.J. 1995. Detection of impact location and magnitude for isotropic plates using neural networks. *Preprint, University of Maryland, Smart Materials and Structures Research Center*, USA.

Kumar M., Found M.S & Howard I.C. 1995. A drop-weight instrumented impact test to compare the effect of single and multiple impacts on CFRP". *Proceedings 2nd International Seminar on Experimental Techniques and Design in Composite Materials, ed. Found M.S., Sheffield Academic Press, Sheffield*: 84-101

Maseras-Guttierez M.A. 1997. Investigation of the impact and the post-impact behaviour of a CFRP with a toughened resin system. *M.Sc. (Res) Thesis, Department of Mechanical Engineering, University of Sheffield*.

Schindler P.M., May R.G., Claus R.O. & Shaw J.K. 1995. Location of impacts on composite panels by embedded fibre optic sensors and neural network processing. *SPIE Proceedings of Smart Sensing, Processing and Instrumentation*: 481-490.

Staszewski W.J., Worden K. & Tomlinson G.R. 1996. Optimal sensor placement for neural network fault diagnosis. *Proc. of ACEDC'96, Plymouth, UK*.

Staszewski W.J. 1996. Optimal sensor distribution for fault diagnosis. Part V: Detection of impacts on a FE composite panel. *Technical Report DRG-BAe-2/96, University of Sheffield, Department of Mechanical Engineering.*

Staszewski W.J. 1996. Optimal sensor distribution for fault diagnosis. Part VI: Detection of impacts in DERA panel using neural networks. *Technical Report DRG-BAe-3/96, University of Sheffield. Department of Mechanical Engineering.*

Staszewski W.J.K., Worden K., Tomlinson G.R. & Ball A. 1997. Optimal sensor locations for impact detection on a composite panel. *In Proceedings of International Workshop on Damage Assessment Using Advanced Signal Processing Procedures - DAMAS 97, Sheffield.*

Weems D., Hahn H.T., Granlund E. & Kim I.G. 1991. Impact detection in composite skin panels using piezoelectric sensors. *47th Annual Forum of the American Helicopter Society, Phoenix, USA.*

133

Section 5: *Modelling*

Experimental Techniques and Design In Composite Materials 4, Found (Ed.)
© 2002 Swets & Zeitlinger, Lisse, ISBN 90 5809 370 0

Damage evaluation in composite structures using modal analysis techniques

A. Nurra & B. Picasso
Department of Mechanical Engineering University of Cagliari, Italy

ABSTRACT: The assessment of failure probability and operational life of composite structures is closely related to the evaluation of the damage produced by static and dynamic loads. The position and intensity of damage selectively influence modal frequencies and modal shapes. Damping is also a fairly good damage indicator since, in relative terms, its variation with damage is more pronounced. This paper tries to give a contribution to the assessment of the potential of modal techniques for non-destructive damage detection in laminated composites. An epoxy-graphite laminate with a $[90_4/45/90_2]_s$ sequence was used to obtain a series of coupon specimens. On each specimen a local damage was artificially induced with various levels of static loading using a three point bending apparatus. Modal analysis was performed prior and after the induction of damage on the cantilever specimen mounted on the testing fixture. A FEM model has been adopted to build a sensitivity matrix that correlates the damage parameters to the variation of the modal frequencies. A computer simulation has been used to validate the algorithms for finding the intensity and position of damage. The relevance of various parameters, such as the number of experimental modes used, and the effect of damping, is analysed.

1 INTRODUCTION

The use of modal analysis techniques to detect and monitor the damage level and position in composite materials, has been proposed by various researchers, (Cawley 1979), (Adams 1975). Damage is present in various forms in composite laminates. Typical situations include matrix cracking, delaminations, fibre debonding, and breakage. The effect of a complex damage state on the modal response is not easily predictable. Some general statements can however be of some help for building a model, in order to establish a relationship between damage and modal response.

a) Matrix cracking on off-axis plies has little effect on the overall stiffness of the laminate. In cross ply laminates a full cracking of the characteristic crack density produces a few percent variation on the laminate bending stiffness.
b) The first natural frequencies show limited variations that sometimes can be attributed to accidental causes, different from damage.
c) Delaminations also produce small effects on the dynamic response of the laminate. The stiffness change is more pronounced in bending.
d) The breakage of fibres has relevant effects on the laminate stiffness. This involves large variations in modal frequencies.

Various authors have investigated the feasibility of using sensible changes in the vibrational parameters to measure the level and localisation of damage. Most of the proposed techniques rely on the measure of the dynamic properties before and after the damage is inflicted. A common assumption for all methods is the presence of linear material behaviour and proportional damping. A basic idea for the localisation of damage is that natural frequencies and modal shapes are differently affected by localised damage. In general, a reduction of stiffness in a localised area, will produce a variation of the elastic energy corresponding to each vibration mode. The corresponding frequency variations will be proportional to this energy change. For cantilever beams tested in bending during flexural vibration, a localised damage will produce larger frequency changes for those modes which have a larger curvature in that damaged zone. Sanders (1979) has used a FEM model in connection with the Talreja (1987) internal state variable theory to develop a system of linear equations, which relate the internal stiffness changes to the variations in the modal response. This paper tries to validate and discuss the limits and potential of this approach. A numerical simulation is made using a FEM model. Finally, a series of experimental tests is performed, with various damage levels inflicted at various positions on composite cantilever specimens. The potential of the method used to detect and measure damage, is discussed and analysed.

2 THEORY

The general theory briefly described in this paragraph, has been developed by Stubbs et al. (1985). The theory includes aspects related to a probabilistic model of the damage detection process and to the most efficient algorithms. The equations of the free vibrations of a discrete system are:

$$\mathbf{M}\,\ddot{\mathbf{x}} + \mathbf{C}\,\dot{\mathbf{x}} + \mathbf{K}\mathbf{x} = 0 \qquad (1)$$

where $\mathbf{K}$, $\mathbf{M}$, $\mathbf{C}$ represent the global stiffness, inertia, and damping matrices of the structure, and $\mathbf{x}$ is the vector of generalised displacements. If the structure undergoes a modification which will involve a corresponding change in the previous matrices, the new dynamic equilibrium equations will be:

$$\mathbf{M}^*\,\ddot{\mathbf{x}} + \mathbf{C}^*\,\dot{\mathbf{x}} + \mathbf{K}^*\,\mathbf{x} = 0 \qquad (2)$$

with:

$$\mathbf{M}^*=\mathbf{M}+\Delta\mathbf{M} \quad \mathbf{K}^*=\mathbf{K}+\Delta\mathbf{K} \quad \mathbf{C}^*=\mathbf{C}+\Delta\mathbf{C} \qquad (3)$$

If, in addition, we assume that the damping matrix is of the proportional type, the equations (1) and (2) can be diagonalised, giving respectively:

$$M_n\,\ddot{y}_n + C_n\,\dot{y}_n + K_n\,y_n = 0$$

$$M^*_n\,\ddot{z}_n + C^*_n\,\dot{z}_n + K^*_n\,z_n = 0$$

$$\text{with} \qquad \mathbf{y} = \begin{Bmatrix} y_1 \\ y_n \\ y_N \end{Bmatrix} = \mathbf{\Phi}\mathbf{x} \qquad (4)$$

and

$$\mathbf{z} = \mathbf{\Phi}^*\mathbf{x}$$

Where M_n, C_n and K_n and M^*_n, C^*_n and K^*_n are modal coefficients given by the expressions:

$$\begin{aligned}
M_n &=\mathbf{\Phi}_n^T\mathbf{M}\mathbf{\Phi}_n & M^*_n &=\mathbf{\Phi}_n^{*T}\mathbf{M}\mathbf{\Phi}_n^* \\
K_n &=\mathbf{\Phi}_n^T\mathbf{K}\mathbf{\Phi}_n & K^*_n &=\mathbf{\Phi}_n^{*T}\mathbf{K}\mathbf{\Phi}_n^* \\
C_n &=\mathbf{\Phi}_n^T\mathbf{C}\mathbf{\Phi}_n & C^*_n &=\mathbf{\Phi}_n^{*T}\mathbf{C}\mathbf{\Phi}_n^*
\end{aligned} \qquad (5)$$

The index n refers to the model number of the structure in free vibration, T means transpose, and * indicates properties after damage has occurred. The variation of the structural properties due to damage is best expressed as a function of the modal parameters:

$$\Delta K_n =\mathbf{\Phi}_n^T\Delta\mathbf{K}\mathbf{\Phi}_n +\Delta\mathbf{\Phi}_n^T\mathbf{K}\mathbf{\Phi}_n +\mathbf{\Phi}_n^T\mathbf{K}\Delta\mathbf{\Phi} +.... \quad (6)$$

where second order terms have been omitted. Similar expressions hold for the mass and damping properties. The n-th damped natural frequency is related to the corresponding undamped frequency by the equation:

$$\omega_{nd}^2 = \omega_n^2\,(1 \text{-} \xi^2) \qquad (7)$$

After damage the relative variations of ω_{nd} are given by the expression:

$$\frac{\Delta\omega_{nd}^2}{\omega_{nd}^2} = \frac{\Delta\omega_n^2}{\omega_n^2} - \frac{\Delta\xi_n^2}{1-\xi_n^2} = \frac{\Delta K_n}{K_n} - \frac{\Delta M_n}{M_n} - \frac{\Delta\xi_n^2}{1-\xi_n^2} \qquad (8)$$

if there is no variation of the mass properties of the structure, the second term in the third member of equation (8) can be discarded, while the first term is given by the expression in (6), discarding the second order terms. Finally we get the following expression which relates the structural damage with the change in natural frequencies and mode shapes:

$$\frac{\mathbf{\Phi}_n^T\Delta\mathbf{K}\mathbf{\Phi}_n +\Delta\mathbf{\Phi}_n^T\mathbf{K}\mathbf{\Phi}_n +\mathbf{\Phi}_n^T\mathbf{K}\Delta\mathbf{\Phi}}{K_n} = \frac{\Delta\omega_n^2}{\omega_n^2} \quad (10)$$

A further modification of the previous equation can be made by putting all the terms which are related to frequencies and mode shapes after damage at the second member:

$$\begin{aligned}
\frac{\mathbf{\Phi}_n^T\Delta\mathbf{K}\,\mathbf{\Phi}_n}{K_n} &= \frac{\Delta\omega_{nd}^2}{\omega_{nd}^2} + \frac{\Delta\xi^2}{1-\xi^2} - \\
&\quad - \frac{\Delta\mathbf{\Phi}_n^T\mathbf{K}\mathbf{\Phi}_n +\mathbf{\Phi}_n^T\mathbf{K}\Delta\mathbf{\Phi}_n}{K_n}
\end{aligned} \qquad (11)$$

The last equation is the basic tool to calculate the stiffness modifications due to damage, provided that the modal shapes and corresponding frequencies before and after damage can be measured. It is, however, rather difficult to measure, with a reasonable accuracy, slight changes in modal shapes. This makes the last term in the previous equation difficult to evaluate. Neglecting this term, will lead to an approximate evaluation of $\Delta\mathbf{K}$. In this way the second member of the equation will contain relative variations of the damped frequencies and damping, the first member, a term containing the undamped mode shapes, and the stiffness matrix variation. During the tests made on composite laminates it has been found that the relative variation of damping is a ratio of two numbers, one close to zero and one close to unity. The second term in equation (11) is thus close to zero, and, for composite materials normally used in engineering applications, it can be neglected. The difficulty in measuring modal shapes with the required accuracy can be overcome if modal shapes are calculated using a finite element model. The number of equations (11) is equal to the number of frequencies which can be measured with the required accuracy.

3 SENSITIVITY MATRIX

The structure taken into consideration is a cantilever beam made of a laminated composite material. Partitioning the beam in a series of finite elements the variation $\Delta\mathbf{K}$ of the stiffness global stiffness matrix can be written as a sum of individual contributions from the singular elements in the form:

$$\Delta \mathbf{K} = \Delta c_1 \mathbf{K}^1 + \Delta c_2 \mathbf{K}^2 + \ldots\ldots + \Delta c_N \mathbf{K}^N$$
$$with \ \Delta c_i = \Delta D^i_{11} \tag{12}$$

the quantities ΔD_{11} are the variations of the flexural rigidities of the various elements modelling a distributed or local damage situation. The corresponding variation of the modal stiffness K_n is given by the expression:

$$\Delta K_n = \boldsymbol{\Phi}^T_n \left[\Delta c_1 \mathbf{K}^1 + \Delta c_2 \mathbf{K}^2 + \ldots\ldots + \Delta c_N \mathbf{K}_n \right] \boldsymbol{\Phi}_n \tag{13}$$

which can be written:

$$\Delta K_n = \alpha_{n1} \Delta c_1 + \alpha_{n2} \Delta c_2 + \ldots\ldots + \alpha_{nN} \Delta c_N = \Delta \omega_n^2 \tag{14}$$

with $\alpha_{ni} = \Phi_n^T \mathbf{K}^i \Phi_n$. The previous equation correlates the change in modal stiffness with the corresponding change in the flexural rigidities of the finite elements. The equation is non-linear because the coefficients α_i depend on modal shapes which, in turn, depend on flexural rigidities. If vibration modes are normalised to give $\Phi^T_n \mathbf{M} \Phi_n = \mathbf{I}$, we have also $K_n = \omega_n^2$ and $\Delta K_n = \Delta \omega_n^2$. The non-linear set of equations (14) has been extensively used by the Authors to detect the stiffness loss, both in intensity and position. It should be observed, that the number of equations (14) equals the number of experimental modes available. This requirement is due to the fact that the matrix α has to be square. While it is desirable to have a large number of finite elements to get a precise description of modal shapes, this number cannot exceed the number of experimental obtained experimentally. This rule can, in some way, be bypassed, if the modal data obtained with a fine mesh are compressed before solving the system (14). We have tried to validate the method exposed using a numerical simulation and the detection of artificially induced damage on composite cantilever specimens.

4 NUMERICAL SIMULATION

A cantilever composite beam with the geometry indicated in Fig. 1 has been considered. The span has been divided in finite elements dividing the beam in five sections. The lamina properties were the following:

$$Ex = 6.5 \ GPa \quad Ey = 150 \ GPa \quad v_{xy} = 0.33$$

The theory of lamination was used to get the value of the flexural rigidity of the laminate. The value obtained was $D_{11} = 10.15$ Nm. Damage was simulated by applying a decrement of D_{11} at particular elements. The damage scenarios for the beam are shown in Table 1.

Table 1. Damage Simulation

	Damage location – Stiffness loss %									
Case	1	2	3	4	5	6	7	8	9	10
1	-	5	-	-	2.5	-	-	-	-	-
2	-	-	-	-	-	-	5	-	-	-
3	-	-	-	30	-	-	-	-	10	-
4	-	10	-	-	20	-	5	-	10	-

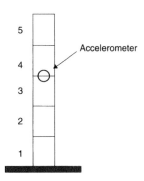

Figure 1. Beam Geometry.

The previous damage scenarios represent a wide range of possibilities for localisation and level of damage. The simulation was based on the following steps:

a) Calculation of the natural frequencies and modal shapes in the undamaged state.
b) Calculation of the matrix α by the system (14).
c) Simulation of damage at selected positions and recalculation of frequencies and modal shapes.
d) Calculation of the vector $\Delta \mathbf{C} = \alpha^{-1} \Delta \omega^2$ of corrections to the element flexural rigidities.
e) Recalculation of the frequencies with the corrected values of the element stiffness.

The process was carried on until the module of the vector $\Delta \omega^2$ was lower than a threshold value. At this point the total variation of the vector $\Delta \mathbf{C}$ would describe the amount and localisation of damage. The process is similar to the piecewise linear approximation in non-linear FEM analysis. The results of the simulation are graphically exposed in Figures 2-5. The number of elements used was 20 for cases 2-4 and 30 for case 1 20 elements in case 1. The accuracy in the detection of the stiffness loss was excellent. It is important to point out that the damage inflicted was in most cases of large magnitude. This enforces the adoption of a non linear approach in the search for the vector Δc. When the vector $\Delta \omega^2$ has a large value it can be convenient to partition the procedure in different steps. The number of elements used in this case seems more than sufficient to achieve an accurate determination of the stiffness change. The time required to converge varies according to the different damage situations. Given the presence of a large number of matrix inversions, the calculation time is strongly dependent on the element number. Using 20 beam elements guarantees that the solution is found in a matter of minutes on a modern PC.

5 EXPERIMENTAL

The specimen geometry, boundary conditions and the position of the static load are shown in Fig. 6. The cantilever specimen was excited to vibrate using an instrumented impact hammer. A low mass accelerometer was

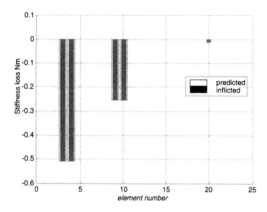

Figure 2. Damage simulation – Case 1.

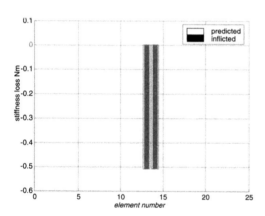

Figure 3. Damage simulation – Case 2.

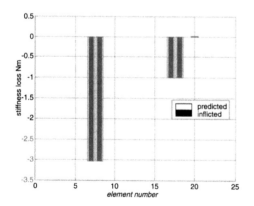

Figure 4. Damage simulation – Case 3.

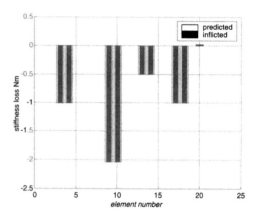

Figure 5. Damage simulation – Case 4.

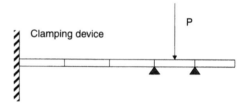

Figure 6. Artificial damage induction by static loading.

mounted on the specimen. The response signals were analysed using a two channel ONO-SOKKI analyser. We have chosen to measure five frequencies and modes which "a posteriori" was found too limited a number. An accurate determination of the modal shapes was impossible with the equipment available. Several factors make this measurement uncertain and inaccurate, such as the repeatability of the excitation on a given point, various non-linearities, and the low magnitude of the differences to be measured. We shell refer in the sequel to modal shapes calculated and not measured. This is one of the major drawbacks in the procedure used, and, perhaps, one of the most important reasons for inaccuracy in the search for the position and level to the damage.

The induction of damage in the specimens was done by means of a bending load applied in the middle of a specimen section. At the ends of the section additional support was provided. Due to the prevalence of the 90° plies, this simple type of loading was apt to produce a rather concentrated state of damage. Two loading levels were applied 145N and 215N respectively. Table 2 shows the damaged sections and the corresponding loads.

The load positions in Table 2 are referred to the five sections in which the beam span has been divided. The loading was applied gradually until the maximum value was reached, and maintained for 30 seconds. From a

Table 2. Load position and level

	Static loads	
	P (N)	Position
1	215	3
2	145	4
3	215	1
4	145	5
5	145	1
	215	4
6	215	3
	145	5
7	215	1
	145	3
8	145	4
	215	5

Figure 7. X-ray image of a specimen after loading.

simple application of the lamination theory it has been verified that the load applied was unable to produce any fibre breakage but simply a concentrated pattern of cracks in the matrix, and (limited) delaminations at the specimen edge. The bending load produced a series of matrix cracks mostly in the tension half of the section while the compression side suffered of a lower, if any, amount of damage. This asymmetry in the damage distribution has been purposely accepted, given the very simple procedure applied to produce the damage. A radiograph of a specimen after three loading levels were applied to contiguous sections is shown in Fig. 7. Matrix cracks and delaminations are clearly visible. Matrix cracks are not always continuous over the specimen width. It is noteworthy that the reduction in stiffness produced by a localised crack pattern like the one shown in Figure 7 is very limited. Stiffness is mostly given by the two 45° layers which are not affected by the damage. Since we have given the priority to the discussion of the potential of detection method, an accurate modelling of the damage situation has been skipped. A model based on internal state variables developed by Talreja (1987) to predict the change in stiffness due to an experimentally detected pattern of matrix cracks, has been applied by Sanders et al. (1992). The experimental procedure for the analysis of the composite specimen was tailored on the need of avoiding accidental disturbances which could influence the modal parameters during the tests. The variation of the first two or three frequencies were of reduced magnitude. The variation of the first frequency is often fractional for the damage levels introduced by the procedure described. For this reason it was decided to keep the specimen in the holding fixture during the static loading. The various phases of the experimental procedure were the following:

a) The specimen is measured and mounted on the holding clamp.

b) A modal analysis is performed. Five frequencies and modal shapes are measured. A zooming procedure is used to measure the frequencies with the best possible accuracy.

c) A localised state of damage is produced on some sections of the specimen by the application of a static loading. In some specimens the section involved is only one, in some others two.

d) A new modal analysis is done on the damaged specimen. The new frequencies are recorded and compared with the undamaged frequencies.

e) The detection algorithm is applied to reveal the level and position of damage.

6 DISCUSSION OF RESULTS

A complete set of results obtained on 8 specimens of approximately the same size is presented in Table 3. The slight differences in the frequency values are due to small changes in specimen dimensions and, perhaps, in the tightening force used to fix the specimen to the clamping fixture. The measured frequencies are immediately referable to ω_{nd}, damped frequencies of the beam. According to equation (11) the measurement of the damping variation for each mode is needed to calculate the undamped frequencies. However the magnitude of the damping values was so small that this correction has been omitted.

In fact the damping value ξ is around 0.005 and the term $\Delta\xi^2/(1-\xi^2)$ is one order of magnitude smaller. This approximation must, however, be removed when large variations of the modal damping are expected. The same line of reasoning holds for the last term in equation (11)

Table 3. Frequency changes caused by damage

	Undamaged	Damaged
	Freq. (Hz)	Freq. (Hz)
Spec. 1		
1	28,6875	28,25
2	189,8438	187,5625
3	540,625	535,225
4	1111,25	1100
5	1727,5	1719,375
Spec. 2		
1	34,313	33,75
2	200,781	200
3	579,375	578,75
4	1138,75	1136,875
5	1805,625	1805
Spec. 3		
1	31,0156	30,8775
2	200,7031	199,922
3	591,406	590,625
4	1170	1169,687
5	1833,125	1828,375
Spec. 4		
1	28,2813	28,2813
2	185,156	185
3	544,531	544,375
4	1085	1083,75
5	1679,69	1676,25
Spec. 5		
1	33,437	33,281
2	199,062	197,812
3	580,937	576,25
4	1135,625	1130,625
5	1815,937	1810,312
Spec. 6	Freq. (Hz)	Freq. (Hz)
1	27,031	27,031
2	175	173,125
3	497,5	496,562
4	1018,125	1010,937
5	1576,562	1573,125
Spec. 7		
1	27,812	27,5
2	177,187	175,937
3	504,375	502,967
4	1025,312	1020
5	1597,187	1588,437
Spec. 8		
1	27,5	27,5
2	177,187	177,187
3	508,75	507,812
4	1025	1022,187
5	1602,87	1598,75

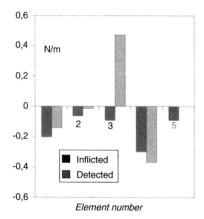

Figure 9. Damage detection – Spec. 5.

which takes into account the variation of the modal shapes with the change in stiffness of the beam material. Figure 9 shows the results of the damage detection method based on the calculation of stiffness variations with five elements. It is apparent that the method fails to provide an accurate localisation of damage. Meaningless, positive, values appear at some locations.

These confusing results were found for all the specimen tested. In trying to give an explanation we have considered that the damage affects only one side of the specimen. Furthermore some finite element calculations have permitted to determine the order of magnitude of the flexural rigidity reduction in the element. This value has been calculated starting from the assumption that the 90° laminas on one side loose half of their original modulus in the direction normal to the fibres during the static loading process. This assumption leads to a value of the stiffness decrement of about 0.6 Nm, which amounts to about 6% of the original bending rigidity. Considering that the modal response is due to the global stiffness and inertia properties of the beam we can conclude that the damage applied has a very limited intensity. The damage is not strictly limited to the section loaded but spread to the contiguous sections. A simple analysis has however evidenced that this effect is not relevant. Another reason for the limited success of the method is due to the fact that five modes are not sufficient to give a full description of the stiffness and inertia properties of the beam. This obligates to limit the number of elements to such a low value with the consequence of an inaccurate determination of the frequencies and modal shapes. The results presented in Figure 9 were obtained using the inverse of the sensitivity matrix α to calculate the vector of the variation of the flexural rigidities of the elements ΔC. The application of the method is straightforward but the algorithm is based on the assumed linearity in the relationship between ΔC and $\Delta \omega^2$. It has been remarked that this linearity is only a rough approximation, to be used when limited variations of ΔC and $\Delta \omega^2$ are expected. We have tried to improve the performance of the algorithm by means of a non-linear optimisation process. The design variables are the components of the vector C, i.e. the flexural rigidities of the elements and the objective function is the function $f(C)=\Delta \omega^2$ which calculates the difference between the present values of ω^2 and the goal values corresponding to the damaged state. When the

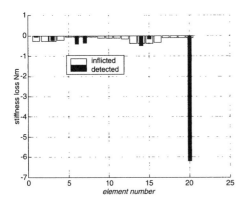

Figure 10. Specimen N. 5 – non-linear method.

goal is reached $f(\mathbf{C}_0)=0$ and the difference between the initial state $\mathbf{C}_{init}$ and the goal state gives the decrement in stiffness at each element. A non linear programming procedure has been used and no convergence problems were experienced. The proposed technique has also been applied to the experimental data. Figure 10 shows the results obtained for specimen n.5. Even if an improvement has been obtained, in comparison with the data obtained from the sensitivity matrix technique, the localisation of damage suffers of large inaccuracies. We believe that the low number of modes used and, perhaps, the influence of the imperfect clamping conditions on the beam could be the possible reasons. A new series of tests is under way to solve these problems and increase the performance of the detection routines.

7 CONCLUSIONS

The aim of this work is to investigate the potentiality of methods based on modal analysis in damage detection for composite materials. The variation of modal parameters with damage is related to the stiffness parameters of the structure with the help of a FEM model. An original technique based on non linear optimisation has been used to search, in the space of the design variables corresponding to the element flexural rigidities, the distribution and localisation of damage. A computer simulation involving several damage situations including various intensities and localisation has been run with success. An experimental validation of the detection techniques has shown the presence of a consistent level of inaccuracy in the determination of the localisation of damage, due, perhaps to the limits and inconsistencies of experimental data. This point will require a further investigation.

The Authors acknowledge the valuable help of Dr. Francesco Aymerich, at the Department of Mechanical Engineering, University of Cagliari, during the non-destructive testing activity.

REFERENCES

Cawley P. and Adams R.D., 1979. The location of defects in structures from measurement of natural frequencies. *J. Strain Analysis*, 14(2): 49-57.

Scott, G. and Scala, C.M. 1982. A review of non-destructive testing of composite materials. *NDT Int.*, 15(2): 75-86.

Adams R.D., Walton D., Filcroft, J.E., and Short, D. 1975. Vibration testing as non-destructive test tool for composite materials. *Comp. Reliability, ASTM STP 580, Amer. Soc. Test. and Mater.* 159-175.

Fox R.L. and Kapoor M.P. 1968. Rates of change of eigenvalues and eigenvectors. AIAA J., 6: 24-26.

Di Benedetto A.T., Gauchel J.V., Thomas R.L. and Barlow J.V. 1972. Nondestructive determination of fatigue crack damage in composites using vibration tests. *J. Mat., JMLSA*, 7(2): 211-215.

Stubbs N. 1985. A general theory of non-destructive damage detection in structures. *Proc. 2nd Int. Symp. on structural control*, Nijhoff, Dordrecht, The Netherlands, 694-713.

Ewings D.J. 1985. Modal Testing: Teory and practice, Research studies, Lechworth, Hertfordshire, England.

Stubbs N. and Osegueda 1987. R. Global non-destructive evaluation of offshore platforms using modal analysis. *Proceedings of the 6th Int. Offshore Mechanics and Arctic Engineering Symposium.* ASME, New York., 2: 517-524.

Sanders D.R., Kim Y.I. and Stibbs N. 1992. Nondestructive evaluation of damage in composite structures using modal parameters. *Experimental Mechanics*, Sept. 1992, 240-251.

Topole K.G. and Stubbs N. 1995. Non-destructive damage evaluation in complex structures from a minimum of modal parameters. *Int. J. Analytical and Experimental Modal Analysis*, 10(2): 95-103.

Talreja R. 1987. Modelling of damage development in composites using internal variable concepts. *Damage Mechanics in Composites, Amer. Soc. Mech. Eng.*, AD-11: 11-16.

Mottersead J.E. 1993. Model updating in structural dynamics: a survey. Journal of Sound and Vibration, 167(2): 347-375.

Experimental Techniques and Design In Composite Materials 4, Found (Ed.)
© 2002 Swets & Zeitlinger, Lisse, ISBN 90 5809 370 0

Investigations on negative Poisson's ratio honeycombs

F. Scarpa
Department of Mechanical Engineering, The University of Sheffield, UK

G. Tomlinson
Department of Mechanical Engineering, The University of Sheffield, UK

ABSTRACT: This paper presents a series of analytical and experimental analysis in order to assess some static and dynamic properties of sandwich structures having a negative Poisson's ratio core.

1 INTRODUCTION

In 1987 an open-celled foam was fabricated that demonstrated a large negative Poisson's ratio (Lakes, 1987). When such a foam is stretched it also expands in the lateral direction. Until then such materials (referred as "auxetic", from the Greek αυξετοσ : to increase) had been thought of as only theoretically possible (Almgren, 1985). In the last years a consistent effort was devoted to the synthesis of polytetrafluoroethylene (PTFE) and microporous ultra-high molecular weight polyethylene (UHMWPE) exhibiting a negative Poisson's ratio behaviour (Evans et al. 1995). A similar effect is also observed on long fibre composites with special stacking layer sequences, and on special honeycombs with re-entrant cell units, which are the subject of this paper.

The potential importance of auxetic materials can be outlined by the interrelationship between the Young's modulus E, the shear modulus G, and the Poisson's ratio ν, for an isotropic material (Young, 1989):

$$G = \frac{E}{2(1 + \nu)} \qquad (1)$$

So, if it is possible to produce a material with a large negative ν, while maintaining the value of E it will have a concomitantly larger shear modulus.

Initial experimental studies on re-entrant cell honeycombs showed the presence of an in-plane negative Poisson's ratio behaviour (Evans et al. 1991). A cellular material theory (Gibson & Ashby, 1997) was implemented in order to describe the anisotropic properties of general honeycombs with auxetic core (Scarpa & Tomlinson, 1998). Starting from this theoretical background the authors have carried out numerical and analytical simulations on static and dynamic properties of sandwich plates with such a particular core (Scarpa & Tomlinson, 1998). The reliability of the theoretical approach used is shown by first experimental works on the subject (Evans et al. 1991), and from further test performed by the authors, which will be shown in the following paragraphs.

2 RE-ENTRANT CELL HONEYCOMBS

Regular honeycombs are made of arrays of hexagonal cells. This specific geometric layout allows one to have isotropic mechanical properties (Gibson & Ashby, 1997). When pulled in one direction the whole honeycomb tends to contract in the transverse one (see Figure 1).

If we have a network composed of re-entrant cell units, as shown in Figure 2, an applied outward stress in one direction induces a global expansion, while a compression in one direction gives rise to a general contraction of the honeycomb in the plane. This kind of behaviour leads to an in-plane negative Poisson's ratio effect.

The mechanical properties of general honeycombs can be described by the geometric layout of the unit cell and the mechanical properties of the core material (Gibson & Ashby, 1997). The geometric parameters which characterise the honeycomb are the internal cell angle θ, the ratio between the cell side α, and the ratio between the thickness of the cell walls and the sides β (Figure 3).

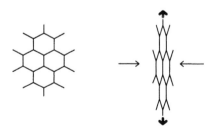

Figure 1. Deformation of a regular honeycomb.

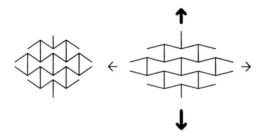

Figure 2. Deformation of a re-entrant cell honeycomb.

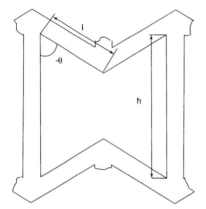

Figure 3. Geometric layout of a re-entrant unit cell.

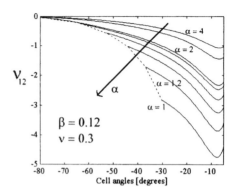

Figure 4. In-plane Poisson's ratio v_{12} for auxetic honeycomb.

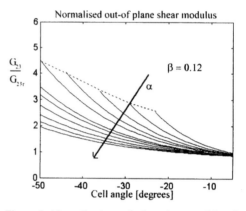

Figure 5. Normalised out-of-plane shear modulus G_{23}.

Assuming the behaviour of the cell walls as one of the Euler-Bernoulli beams, and the imposed periodic boundary conditions to the cells, it is therefore possible to compute the mechanical properties of a general honeycomb (Gibson & Ashby, 1997). Further refinement can be obtained taking into account the axial deformation in the plane and shear correction factors for thick walls ($\beta > 0.25$).

The most important conclusion that can be drawn from the above theory is the fact that general honeycombs have orthotropic mechanical properties, as can be stated from the following relationships between in-plane Young's moduli and Poisson's ratio:

$$E_1 v_{21} = E_2 v_{12} \qquad (1)$$

Another significant result is the relative independence of the in-plane Poisson's ratios from the mechanical properties of the core material, as it can be stated from the equation describing the Poisson's ratio v_{12}:

$$v_{12} = \frac{\cos^2 \theta}{(\alpha + \sin \theta)\sin \theta} \frac{1 + (1.4 + 1.5 \cdot v)\beta^2}{1 + (2.4 + 1.5 \cdot v + \cot^2 \theta)\beta^2} \qquad (2)$$

From Equation 2 it can be observed that a negative cell angle (i.e., a re-entrant cell geometry) allows a negative Poisson's ratio value (see Figure 4). Its magnitude

is also dependent on the side ratio α. Due to the re-entrant topology of the cell, the internal cell angle cannot overcome a maximum value to avoid the touch of the opposite vertex during the deformation (Scarpa & Tomlinson, 1998):

$$\theta_{max} = \cos^{-1}\left(\alpha/2\right) \qquad (3)$$

The presence of the Poisson's ratio core value v is significant only for high values of thickness ratio β, and is generally neglectable for the other cases (Gibson & Ashby, 1997).

One important mechanical property for sandwich structures applications is the through-the-thickness shear modulus, which assess the shear resistance of the laminate. Figure 5 shows the behaviour of the shear modulus G_{23} versus the internal cell angle and a parameter. The values are normalised with the ones of a corresponding regular honeycomb ($\alpha = 1$, $\theta = 30°$) with the same core material. For various combinations of geometric parameters, one can obtain a negative Poisson's ratio honeycomb with enhanced out-of-plane shear moduli up to 4 times compared to the one of a corresponding regular honeycomb.

3 STATIC AND DYNAMIC PERFORMANCES OF SANDWICH PLATES

Due the orthotropic properties of the auxetic honeycomb, is it possible to use the formulations related to general orthotropic sandwich panels (Whitney, 1987). The authors have analysed general plates both from the analytical and numerical point of view (Scarpa & Tomlinson, 1998). The cylindrical bending case can offer a first assessment of the mechanical performances of sandwich structures with such core material (Pagano, 1969). If we assume isotropic face sheets and neglecting the rotary inertia terms, the equations of motion for an infinite sandwich plate in the y-direction are the following (Whitney, 1987):

$$\begin{cases} D_{11} \dfrac{\partial^2 \psi_x}{\partial x^2} - G_{13}h\left(\psi_x + \dfrac{\partial w}{\partial x}\right) = 0 \\ G_{13}h\left(\dfrac{\partial \psi_x}{\partial x} + \dfrac{\partial^2 w}{\partial x^2}\right) + N_x^i \dfrac{\partial^2 w}{\partial x^2} + q = \rho \dfrac{\partial^2 w}{\partial t^2} \end{cases} \quad (4)$$

The term D_{11} represents the reduced bending stiffness term, while G_{13} is the out-of-plane shear modulus. The generalised co-ordinates are the normal displacement w and the rotation ψ_x around the x-axis. N_x^i and q are the in-plane compression and the uniform distributed load respectively. Making use of the simply supported boundary conditions (Whitney, 1987) one can derive analytical solutions for the maximum static displacement, buckling load and natural frequency of the sandwich plate.

Figure 6 shows the maximum displacement for a unit distributed load normalised to the analogous result for a sandwich plate with regular honeycomb core. It can be observed that for low side cell ratios ($\alpha < 2$) and lower internal cell angles, the maximum displacements are less than the ones of a corresponding hexagonal honeycomb (ratio lower than 1), leading to the conclusion that in this parameter range the laminate has a greater static bending stiffness. For higher α values the maximum displacements of the auxetic laminate are higher compared to the ones of a regular sandwich laminate. This behaviour can be explained by the fact that the shear modulus G_{13} is significantly higher at low α's and θ's, and it affects the denominator of the following equation describing the maximum displacement (Pagano, 1987):

$$w_{max} = \frac{qa}{48D_{11}}\left[\frac{6aD_{11}}{G_{13}h} + \frac{5}{8}a^3\right] \quad (5)$$

Figure 7 shows the variation of the critical buckling load versus the geometrical parameters of the honeycomb. Small negative cell angles with low α ratios cause an increase of the buckling load (up to 17%) compared to a similar structure with a regular core. Once again the out-of-plane shear modulus affects greatly the behaviour of the critical buckling load as it can be seen from the following equation:

$$N_{cr} = \frac{G_{13}D_{11}h\pi^2}{D_{11}\pi^2 + G_{13}ha^2} \quad (6)$$

For the case of free vibrations in the range of geometric parameter that gives rise to an increase of the static performances of the plate, the increase of the relative density allows a lowering of the fundamental frequency compared to the one of a sandwich plate with regular core, as seen in Figure 8.

In fact, both the density and the shear modulus scale as β and have an increase in the same range of geometric parameters (Gibson & Ashby, 1997). The fundamental frequency is inversely dependent on the product ρG_{13}, and this fact allows a decrease of the frequency in the above range of α's and θ's:

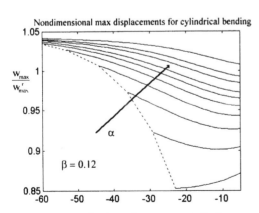

Figure 6. Normalised maximum central displacement for cylindrical bending.

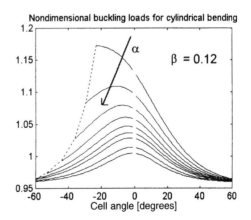

Figure 7. Nondimensional buckling load for cylindrical bending.

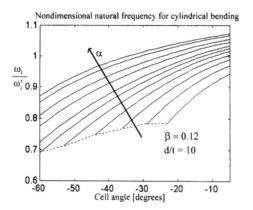

Figure 8. Normalised fundamental frequency for cylindrical bending.

$$\omega_1 = \frac{\pi^2}{a^2} \sqrt{\frac{G_{13} D_{11} a^2}{\rho\left(G_{13} h a^2 + D_{11} \pi^2\right)}} \qquad (7)$$

For higher α's ($a > 2$) and low θ's there is an enhancement of the fundamental frequency, mainly due to the decrease of the relative density in the same interval.

4 EXPERIMENTAL TESTS

So far only few results are available in order to characterise from the experimental point of view the mechanical properties of auxetic honeycombs. First tests were carried on using a photographic system in order to detect the magnitudes of the synclastic curvatures of aluminium and paper re-entrant cell honeycombs (Evans et al. 1991).

One important issue is the scarcity of samples of auxetic honeycomb available. In order to verify the predictive values of the Cellular Material Theory (Gibson & Ashby, 1997), we used a Nomex™ sample produced by Hexcel Composite Ltd, Duxford, and other samples with different unit cell layout made by a stereolythography process (IMC Nottingham). The Nomex™ sample was analysed via an Image Data Processing unit while subjected to a uniform displacement field, in order to calculate the deformations and to verify the in-plane values of Poisson's ratios. Modal tests were carried out on the SL samples in order to detect the free-free natural frequencies of the plates. The results were compared with the ones given by a Finite Element analysis carried out on a commercial code.

4.1 *Image Data Processing*

A Nomex™ sample of dimensions 25 x 26 cm with parameters $\alpha = 2$, $-\theta = -30°$ and $\beta = 0.017$ (Figure 8) was inserted into a frame and loaded at one and by a superposition of calibrated weights of 227g, 454g and

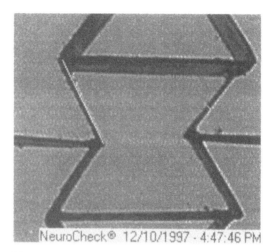

Figure 9. Unit cell of the Nomex™ sample.

680g. The above weights were placed in a bar in order to impose a displacement field at one end of the sample. To assure a better uniformity on the displacements, a vulcanised rubber layer was placed between the bar and the sample. Special care was taken to insure the colinearity of the weights and the general contact conditions between frame and sample, in order to simulate as much as possible a plain strain condition. The tests were performed at an average temperature of 17°C.

For each loading condition four test repetitions were carried out. After each loading, an interval of 15 seconds was observed to avoid hysteresis effects on the loading conditions. While the sample was loaded, a Sharp S300 videocamera connected to a proprietary acquisition card managed by the Neurocheck® Image Data software recorded images at 15 frames per second rate. Such camera dynamics was sufficient in order to detect the quasistatic deformations of the honeycomb. Special care was taken to calibrate the reference images in order to assign correct lengths per pixel via an integrated routine. From this point of view light conditions are an important parameter to be controlled. The frame was inserted in a semi-closed box provided with a uniform emission light. This feature allowed to reduce the number of calibrations required in order to identify a sufficient number of unit cells to be taken as a reference frame. Six cells near the central region were chosen, so as to identify a zone where the plain strain conditions should have been satisfied in a more acceptable way.

The strains in the two directions were computed using the following formulations:

$$\varepsilon_x = \frac{\Delta X}{X_{ref}}$$
$$\varepsilon_y = \frac{\Delta Y}{Y_{ref}} \qquad (8)$$

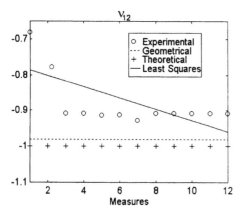

Figure 10. Experimental ν_{xy} Poisson's ratio for Nomex™ sample.

Figure 11. SL auxetic samples.

The global result was computed averaging the strains spotted for each cell. Figure 10 shows the computed Poisson's ratio $\nu_{xy} = \dfrac{\varepsilon_x}{\varepsilon_y}$ due to a uniform displacement field in the y-direction.

The average Poisson's ratio calculated on 92% of the results was about the value of -0.932, close to the theoretical one of -1 for an infinite honeycomb and -0.998 taking into account axial deformations and shear effects (Equation 2). This leads to a relative error of the 6.6%, which is in line with the experimental results carried by Evans (Evans et al. 1991).

4.2 Modal tests

Stereolythography samples of various side wall aspects and internal cell angles were used to conduct modal analysis tests. The target was to identify the vibrational behaviour of thick orthotropic plates with a core material more rigid than the one provided by aramid paper, as used in the previous test. Cast epoxy resin allows a general stiffer structure with a sufficient weight. In this paper we present the results related to a sample with $\alpha = 4$, $\theta = -30°$ and $\beta = 0.38$, as shown in the left side of Figure 11. The size of the samples was 0.255 x 0.278 x 0.04 m.

The sample was suspended freely by elastic bands connected to a rigid frame. An electromagnetic shaker was connected to the sample via a push-rod stinger in order to excite with a broadband white noise between 0 and 1024 Hz. A force captor PCB 086C03 was connected to a passing rod fixed with washers to the sample. This layout was useful to guarantee an higher energy input to all the zones of the samples, and to avoid the energy losses due to the high damping of the cast epoxy. Special care was taken in the placement of the input device, in order to avoid as much as possible localised stiffness and unexpected rotary inertia and rigid body mode effects.

A B&K 4393 accelerometer was placed in 13 different locations in order to obtain sufficient Frequency Response Functions to synthesise the first 4 modes. The small size of the sensor allowed the positioning in the small areas constituted by the crossing of the unit cells.

The accelerometer, the force transducer and the shaker were connected through a B&K acquisition front-end to a HP 715 Unix workstation of 64 MB of RAM hosting a modal analysis software LMS Release 3.4a. The MIMO module of the program was used to detect the FRF's and match them with a geometry mesh already created with the geometry module. Through the MIMO module both the sensors and the excitation signals were handled through a guided interface program. The signals from the sensors were Fourier transformed with a sampling rate of 2048 HZ, so respecting the Nyquist criteria. A Hanning windowing on the signals was performed, and a 70% of overlap used. 60 averages were performed before the final acquisition of each FRF. H1 estimators were adopted in order to assess coherence functions.

The calculated FRF were then associated to a mesh representing the tested sample and synthesised with SDOF Circle Fitting and MDOF Frequency Domain Methods (Maia et al. 1997). The two models were then compared with a MAC criteria in order to check common high correlation rates between the results.

Finite Elements were used to relate the experimental results. The commercial code ANSYS53 running on a SG multiprocessor workstation with 1 GB of RAM was utilised. The model consisted of 578 brick elements SOLID64 with the capability of representing general anisotropic materials. The choice of the brick elements, instead of shell ones was explained by the thickness of the samples, which raised the 18% of the minimum side length of the samples. The mechanical properties were calculated using the Cellular Material Theory, taking into account in-plane axial deformations and shear correction factors. The natural frequency and modes were calculated using a Subspace Iteration routine, and the first 15 modes were extracted. The comparison of the results is given in Table 1.

Table 1. Comparison between FEM and experimental natural frequencies

Mode	FEM	Test	Error %
torsional	70.47	71.27	1.12
shear	294.2	285.8	-2.9
flexural	314	334.4	6.1
flexural-torsional	372	398.4	6.6

As it can be observed, there is a general good agreement between numerical and experimental results. The first two modes are not flexural for the low torsion rigidity and high $G_{23}h$ product of the plate. In the case of the flexural and combined torsional-flexural modes, the effects of the connecting rod to the force captor are more visible, leading to higher participation of the rigid body modes. This introduces a wider error in the identification of the modal properties, a situation which is cumbersome to solve for the particular layout of the sample, where only a very small area is available to attach sensors of normal usage. Nevertheless, the relative error is still acceptable, with a peak of 6.6%.

5 CONCLUSIONS

Auxetic honeycombs can be applied in various structural applications. Their enhanced mechanical properties can be used to design sandwich structures with increased bending stiffness, augmented critical buckling loads, and more generally on structures customised for special purposes, due the variability of their mechanical properties due the geometry of the unit cell. The authors have previously investigated possible applications for sound radiation insulation (Scarpa & Tomlinson, 1998).

First experimental results seem to assess the reliability of the CMT of Gibson & Ashby to describe the mechanical properties of these honeycombs.

REFERENCES

Almgren R.F. 1985. An Isotropic Three-Dimensional Structure with Poisson's ratio = -1, *J. Elasticity* 15: 427-430.

Caddock B.D., Evans K.E. & Masterws I.G. 1991. Honeycomb cores with a negative Poisson's ratio for use in sandwich panels. *Proceedings ICCM/8* edited by S.W. Tsai and G.S. Springer, SAMPE, Covina, CA.

Evans K.E. 1995. Microstructural modelling of auxetic microporous polymers. *J. Mat. Sci.* 30: 3319-3332.

Gibson L.J. & Ashby M.F. 1997. *Cellular Solids*. Cambridge. Cambridge Solid State Science Series.

Lakes R. 1987. Foam Structures with a Negative Poisson's Ratio. *Science* 235: 1038-1040.

Maia N.M.M. & Silva J.M.M. 1997. *Theoretical and Experimental Modal Analysis*. Taunton, Somerset. RSP Ltd.

Pagano N.J. 1969. Exact Solutions for Composite Laminates in Cylindrical Bending. *J. of Comp. Mat.* 3: 398-411.

Scarpa F. & Tomlinson G. 1998. On Static and Dynamic Design Criteria of Sandwich Plate Structures with a Negative Poisson's ratio Core. *Proceedings of the 4th European and 2nd MIMR*, 559-566. Harrogate, UK.

Scarpa F. & Tomlinson G. 1998. Vibro-acoustics and damping analysis of negative Poisson's ratio honeycombs. *Proceedings of Smart Structures and Materials*, 345-357. S. Diego, CA.

Whitney J.M. 1987. *Structural Analysis of Laminated Anisotropic Plates*, Lancaster: Technomic Publishing Company, Inc.

Young W. 1989. *Roark's Formulas for Stress & Strain - 6th Edition*. New York: McGraw-Hill.

Experimental Techniques and Design In Composite Materials 4, Found (Ed.)
© 2002 Swets & Zeitlinger, Lisse, ISBN 90 5809 370 0

Influence of Scale Effects on the Tensile Strength of Notched CFRP-Laminates

M. Luke
Deutsche Bahn AG, formerly Gerhard Mercator University of Duisburg, Germany

R. Marissen
Delft University of Technology, The Netherlands

H. Nowack
Gerhard Mercator University of Duisburg, Germany

ABSTRACT: Composite materials are most promising candidates for light weight structures in railway applications. However, in order to take advantage of the entire weight reduction potential the designer has to take into consideration through-thickness stress concentrations and damage mechanisms due to the presence of notches and other geometrical discontinuities. The objective of the present study is to introduce an experimentally and numerically based approach to quantify the tensile strength reduction of $[0_2,45_2,-45_2,0]_s$ CFRP-laminates due to scale effects.

1 INTRODUCTION

A well known problem in designing structures is to distinguish whether the available input data describes the material behaviour itself or is already dependent on the specimen geometry or the test method used to derive the data. Especially inhomogeneous materials such as composite materials which already represent complex structures themselves tend to be sensitive to scale effects.

In Figure 1 an example for a basic scaling problem is depicted. The tensile strength of $[0_2,45_2,-45_2,0]_s$ T300/PEI test coupons with double edge notches and an equal-sized thickness, t, is plotted against the specimen width. Although the notch diameter to specimen width ratio, 2a/w, was kept constant the tensile strength exhibits a significant drop as the specimen width is increased. Compared to a tensile strength of 444MPa, based on the gross section of the specimen with a width of 9mm, a strength reduction of approximately 31% (307MPa) for specimen with a width of 45mm was observed. For specimen widths larger than 45mm the tensile strength varies around a mean value of 307MPa.

In separately conducted tensile strength tests with unnotched specimen with widths up to 60mm an analogous tensile strength reduction could not be observed. The variations of the measured strength values of the unnotched specimen were rather similar to those of the notched specimen with widths larger than 45mm.

Some practical models considering scale effects have already been proposed to predict the tensile strength of notched laminates. Whitney & Nuismer (1974) introduced a semi-empirical concept comparing analytically obtained stress values at a characteristical distance from the notch root. Kortschot et al. (1991) derived a strength prediction model of cross-ply laminates taking into account the damage development just before fracture (terminal damage state). Shahid et al. (1995) developed

an accumulative damage model based on a finite element approach in which local damage processes during the load application are considered. Marissen et al. (1995) introduced the ineffective length model for tensile strength predictions of woven fabrics which takes into consideration material data as well as microstructural characteristics.

However, it needs additional work to improve the physical understanding of the complex stress/strain behaviour of notched composites. Tensile tests which were previously performed (Luke et al. 1993) on notched laminates with the same stacking sequence, the same fibers, but different matrices indicated that the load transfer by the matrix between the individual layers may be of predominant importance for the tensile strength of notched carbon fiber reinforced plastics (CFRP). For this reason the load transfer was investigated in more detail in the present study.

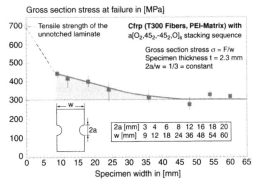

Figure 1. Experimentally determined tensile strength of a $[0_2,45_2,-45_2,0]_s$ T300/PEI laminate as a function of the specimen width for a constant ratio of 2a/w.

100 MPa = 27% σ_f 150 MPa = 40% σ_f

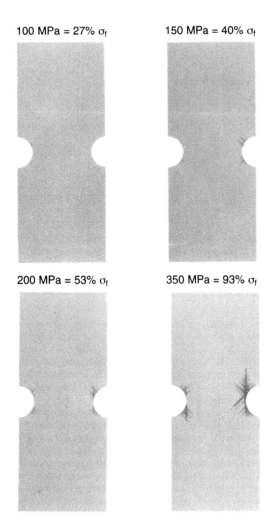

200 MPa = 53% σ_f 350 MPa = 93% σ_f

Figure 2. X-ray radiographs of the damage development of a stepwise tensile loaded $[0_2,45_2,-45_2,0]_s$ T300/PEI laminate.

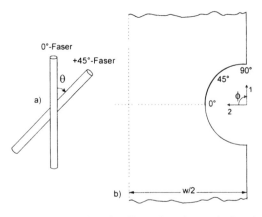

Figure 3. Definition of a) fiber orientation angle Θ and b) notch angle ϕ.

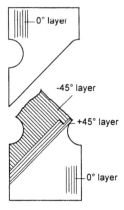

Figure 4. Schematic representation of the characteristical fracture mode.

2 EXPERIMENTAL RESULTS

2.1 *Phenomenological damage development*

In order to study the complex stress/strain behaviour of the tensile loaded $[0_2,45_2,-45_2,0]_s$ T300/PEI-laminates in the vicinity of the notch, a detailed X-ray analysis of the damage development has been performed. Figure 2 shows a series of X-ray radiographs of a specimen with a width of 18mm. The tests have been interrupted at 27%, 40%, 53% and 93% of the tensile strength value, anticipated from previous tests. To facilitate the stress transfer around the notches into the load-carrying fibers a stacking sequence containing a high percentage of ±45°-layers had been choosen.

Consequently ±45° matrix cracking is initiated first (100 MPa) at notch angles between ϕ=+37.5° and ϕ=+42.5° and between ϕ=-37.5° and ϕ=-42.5° respectively. A definition of the fiber orientation angle θ and the notch angle ϕ is given in Figure 3. With increasing loads the number and length of the ±45° matrix cracks increases until 0° matrix cracks, emanating from the notch root, can be observed at a load of 200MPa. The damage state at 350MPa is characterized by a high density of ±45° matrix cracks over a notch angle range of ϕ=0° and ϕ=±60° as well as asymmetrically grown 0° matrix cracks with a length within the order of the notch diameter 2a.

The damage development as well as the finally reached damage state are characteristical for the laminate investigated, independently of the specimen width. In addition, all specimen also fractured in the same characteristical way. Figure 4 shows a schematic representation of a notched tensile loaded $[0_2,45_2,-45_2,0]_s$

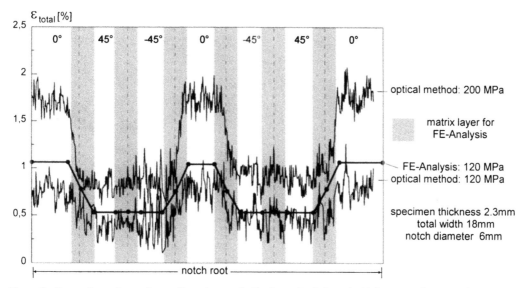

Figure 5. Comparison of experimentally and numerically determined through-thickness notch root strains, ε_{total}, at different nominal stresses.

T300/PEI-laminate after being fractured. It can be seen that failure occurred under a +45° or -45° orientation. It is important to note that all load-carrying 0°-fibers fractured according to the ±45° orientation of adjacent ±45°-layers. The 0°-fibers of the surface layers fractured under +45° whereas the 0°-layers in the centre of the laminate fractured under -45°.

2.2 Notch root deformation measurements

To evaluate the matrix load transfer between individual laminate layers the deformations at the notch root of the specimens across the thickness were measured using an optical technique. Two parallel golden marker lines were applied across the notch root by physical vapour deposition. The distance between the two marker lines, l_0, was 1mm. While loading the specimen the displacement, Δl, of the marker lines was monitored by a 1024x1024 pixel CCD video camera. The strain at the notch root, ε_{total}, was calculated at selected load levels by relating the measured displacements to l_0 (for more experimental details see Luke 1998).

In Figure 5 the magnitude of ε_{total} across the notch root is shown as a function of the specimen thickness. It can be seen that the ε_{total}-values of the individual 0°-layers exceed those of the ±45°-layers at all load levels. This finding may be explained by the fact that ε_{total} is a superposition of fiber and matrix normal- and shear-deformations which occur between the marker lines (Fig. 6). For the 0°-layer the contribution of the shear deformation of the matrix is larger than for the ±45°-layers which leads to higher ε_{total}-values. The continuous measurement of the deformation across the specimen thickness revealed deformation gradients

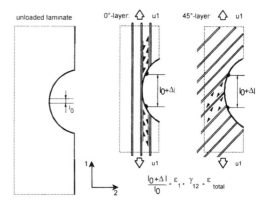

Figure 6. Influence of matrix shear deformation on ε_{total} of 0° and 45° laminate layers.

between 0°- and ±45°-layers which are characteristical for the load transfer of the matrix between the layers. How strong the gradients are, depends on the loading. They increase with increasing load until matrix cracks are formed. The cracks lead to discontinuities in the deformation distribution (Fig. 5, 200 MPa). To enable a crack free load transfer over a large load range a thermoplastic PEI (Polyetherimide) matrix with a fracture strain up to 60% was selected.

3 FE-ANALYSIS

For a further, more detailed investigation of the notch stresses and strains and the load transfer through the

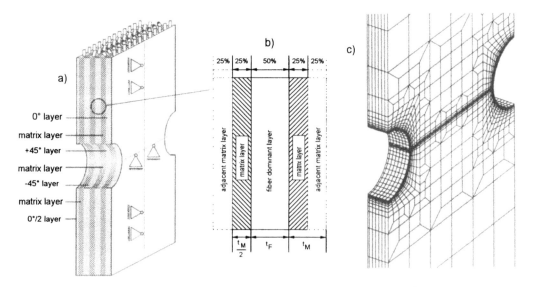

Figure 7. (a) Finite element model of the notched multilayer composite, (b) subdivision of fiber/matrix layers of the FE model, (c) FE mesh used.

matrix a special three dimensional finite element analysis was performed. Each $0°$ or $±45°$ fiber/matrix layer of the laminate was represented by a fiber dominant layer sandwiched between two coupling layers of 100% matrix material. Because the experimentally determined fiber volume fraction of the material was approximately 50%, the thickness of the two outer 100% matrix layers was chosen as 25% of the actual fiber/matrix layer thickness. The thickness of the fiber dominant layer in the middle was 50% as well (Fig. 7b).

The elastic constants of the fiber dominant layers were adjusted such that the global stiffness of the original laminate was obtained. In order to reduce the number of elements adjacent matrix layers were merged and the surface matrix layer was neglected. Because of the asymmetrical behaviour of the $±45°$-layers half of the laminate with respect to the laminate midplane had to be modelled. All calculations were performed linear-elastically using the same mesh (Fig. 7c).

The calculated distribution of the deformations across the thickness is also included in Figure 5. It can be seen that the finite element results led to a stress/strain behaviour at the notch root which is analogous to the experimental findings as long as no matrix cracking has occurred. (By a fitting of model parameters such as, the matrix stiffness or the thickness of the matrix coupling layers even better agreement can be achieved).

The three dimensional finite element analysis was performed to investigate the differences in the stress/strain behaviour of the individual laminate layers which arise because the layer constraints develop in a different manner. For all models the $(2a/w)$-ratio was 1/3 and the thickness, t, was equal-sized.

Figure 8 shows the σ_x-stresses acting in fiber directions within the $+45°$-layers near the notch for different specimen widths at a gross section stress of 120 MPa. All fiber stresses were normalized with respect to the fiber failure strength $\sigma_{f,fiber}$ of 3530 MPa. At an angle measured from the centre of the notch of $\phi=45°$ maximum stress values between 14% to 22% $\sigma_{f,fiber}$ were reached. The $+45°$ fiber stresses calculated in models having a larger width were higher than those calculated in models having a smaller width. This most interesting result was also found for the $-45°$-layers at $\phi=-45°$.

Normal stresses are introduced into $45°$ fibers by means of shear stresses. The stress magnitude of the normal stresses which are introduced depends on the absolute fiber length. Generally a minimum length of the fiber is required. An optimum of stresses can be transferred to the fiber as soon as a sufficient fiber length is reached. This explains why higher stresses can be transferred to $45°$ fibers as calculated in the models with a larger width. Analogous stress transfer effects have already been discussed by Marissen et al. (1995) in conjunction with their ineffective length model.

4 DISCUSSION

Examinations of fractured specimens confirmed that the stress/strain behaviour of the $±45°$-layers is indeed important for the failure of the laminates. As stated above in all cases fracture of the load-carrying $0°$ fibers occurred along positions which were prescribed by adjacent $±45°$ fibers emanating from the notch at an angle ϕ of $±45°$ (Fig. 4).

Moreover, the X-ray radiographs of the T300/PEI

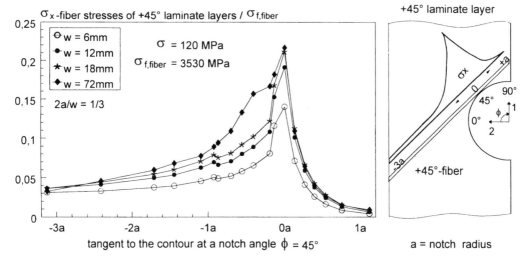

Figure 8. Comparison of σ_x-fiber stresses in the +45° laminate layers of the FE-models with different widths.

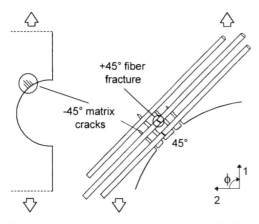

Figure 9. Schematic representation of +45° fiber fracture due to -45° matrix cracks.

a) 0° fiber fracture due to +45° matrix cracks

b) laminate failure due to 0° fiber fracture

Figure 10. Schematic representation of the characteristic tension failure mode of a $[0_2,45_2,-45_2,0]_s$ T300/PEI-laminate.

laminate in Figure 2 showed that the damage development is concentrated on two different regions along the notch contour. The matrix cracking starts first in the ±45°-layers at a notch angle ϕ of ±45°. Later on the 0° matrix cracks begin to form, emanating from the notch root.

However, 0° matrix cracking leads to a release and redistribution of stresses at the notch root. Consequently the stress concentration at the notch root of 0°-layers may no longer be as important for the fracture of the 0° fibers and with that for the failure of the entire laminate.

The stress concentrations in the ±45°-layers cannot be released due to matrix cracking. The failure of the laminate is initiated by a fracture of ±45° fibers due to

matrix cracks in the adjacent ∓45°-layers. The matrix cracks represent sharp notches adding additional stresses to the strained ±45° fibers (Fig. 9). A fiber fracture mechanism like this has also been observed by Jamison et al. (1984) at laminates with different stacking sequences. The loads which are no longer carried by the fractured ±45° fibers are transferred to adjacent 0°-layers and lead to 0° fiber fracture and the failure of the entire laminate (Fig. 10a, b).

From these considerations the conclusion can be drawn that the measured decrease in tensile strength of specimen with a larger width is predominately dependent on the increase of the ±45° fiber stresses in the vicinity of the notch.

5 CONCLUSIONS

1. Experimental and numerical investigations showed that the deformations at the notch root varied from layer to layer across the laminate thickness depending on the fiber orientation.

2. A mesostructural three dimensional finite element analysis of the notch problem provided important new understanding with respect to the critical load transfer mechanism.

3. The tensile strength reduction (scale effect) of the laminate investigated is caused by differently loaded ±45° fibers in the vicinity of the notch.

4. Increasing the tensile strength of the critical ±45° fibers helps to take full advantage of the 0° fiber strength and therefore improves the tensile strength of the whole laminate.

6 ACKNOWLEDGEMENT

The authors wish to thank Ekkehard Maldfeld for his help with the finite element calculations and Prof. Gutjahr for his help with the optical strain measurements in the notch root.

REFERENCES

Jamison R.D., Schulte K., Reifsnider K.L. & Stinchcomb W.W. 1984. Characterization and Analysis of Damage Mechanisms in Tension-Tension Fatigue of Graphite/Epoxy Laminates. *Composite Materials: Effect of Defects in Composite Materials, ASTM STP* 836: 21-55. Philadelphia: ASTM.

Kortschot M.T., Beaumont P.W.R. & Ashby M.F. 1991. Damage Mechanics of Composite Materials: III - Prediction of Damage Growth and Notched Strength. *Composites Science and Technology.* 40: 147-165. Amsterdam: Elsevier.

Luke M., Schulte K. & Nowack H. 1993. Monotonic and Cyclic Behavior of Notched CFRP-Components and Prediction. In *Proc. of the 5th Int. Conf. on Fatigue and Fatigue Thresholds*, Montreal, 3-7 May 1993: 477-483. Engineering Materials Advisory Services LTD.

Luke M. 1998. Experimentelle und numerische Beanspruchungsanalyse an gekerbten CFK-Laminaten unter Zugbeanspruchung. *Ph.D. Thesis, Gerhard Mercator University of Duisburg.* Aachen: Shaker Verlag.

Marissen, R., Brouwer R. & Linsen J. 1995. Notched Strength of Thermoplastic Woven Fabric Composites. *Journal of Composite Materials.* 29: 1544-1564. Technomic.

Shahid I., Sun H.-T. & Chang F.-K. 1995. Predicting Scaling Effect on the Notched Strength of Prepreg and Fiber Tow-Placed Laminated Composites. *Journal of Composite Materials*, 29: 1063-1095. Technomic.

Whitney J.M. & Nuismer R.J. 1974. Stress Fracture Criteria for Laminated Composites Containing Stress Concentrations. *Journal of Composite Materials*: (12): 148-160. Technomic.

Experimental Techniques and Design In Composite Materials 4, Found (Ed.)
© 2002 Swets & Zeitlinger, Lisse, ISBN 90 5809 370 0

Examination of the strain transfer efficiency in composites with a descrete interphase region

S.A. Hayes, R. Lane & F.R. Jones
Dept. Engineering Materials, University of Sheffield, Sheffield, S1 3JD, England

ABSTRACT: Although many micromechanical models for the stress transfer between matrix and fibre have been proposed, little work has taken into account the existence of the interphase. The presence of such a region is inevitable due to the nature of composites and their manufacture and as such it is vital to establish its effect on the stress transfer between matrix and fibre. This study examines the relationship between strain transfer characteristics and the properties of the interphase region using finite element analysis.

Two parameters were defined to describe the behaviour of these model systems; a Stress Transfer Efficiency (STE) and a Stress Transfer Length (STL). The STE was shown to be matrix property sensitive whilst the STL was sensitive to the properties of the interphase.

Plasticity within the matrix and interphase was shown to have a large bearing on the development of strain in the embedded fibre.

1 INTRODUCTION

Fibre reinforced polymer matrix composites rely on bonding between the reinforcing fibres and matrix resin to give them their unique combination of properties. Understanding the matrix to fibre stress transfer mechanism is vital to understanding composite behaviour and performance. Strain transfer will inevitably occur through an interphase between the fibres and the matrix so this region is of obvious importance to the performance of a composite. To do this the micro-mechanics of the composites have to be scrutinised because they offer the key to how micro-phenomena influence macro properties.

It is well accepted that the region between fibre and matrix in composite materials is neither distinct nor perfect and has been described as an interphase [1,2]. The interphase itself is known to be complex even though it accounts for only a fraction of the total composite weight. It may be a diffusion zone, a nucleation zone, a chemical reaction zone or any combination of these [3]. As a result, the interphase properties are likely to be inhomogeneous through its thickness due to a graduation in both composition and morphology [4]. A description of the interphase is also unclear, but a sensible definition is provided by Drzal [2] who states that the boundaries are the 2D fibre/matrix interface and the point in the matrix where the properties do not deviate from those of the bulk matrix. An interphase will inevitably be present in commercial composites because of the use of fibre sizing resins. Often the composition

of these sizings is unknown, which leads to further complications when trying to correlate composite properties to micro-phenomena in research studies. The yield of the matrix and also the interphase are frequently neglected in analyses.

A study of the effect of the interphase on strain transfer has been carried out [5]. This consisted of a modification to the shear-lag theory to include the interphase which produced a prediction of its shear modulus and thickness. An iterative calculation was then carried out until the solution reached reasonable agreement with prior experimental data, so determining the interphase properties. As a result the accuracy of these 'properties' may be questioned. The study did however present some interesting ideas. It was proposed that debonding could not occur at the fibre/ matrix interface, but within the interphase itself by shear, reiterating the importance of plasticity in the analysis. It was also surmised that an enhanced stress concentration increased the likelihood of debonding. It should be noted that stress concentrations increase with a stiffer or thinner interphase. The results were also correlated to an FEA study of the fibre pull-out test. It was shown that the interphase shear strength and post debonding coefficient of friction varied with interphase thickness, with a weak interface being produced by a low shear strength interphase.

It can be seen that the contribution of the interphase to performance, strength, response and the failure modes in composites is potentially very important, and should therefore not be overlooked in analytical modelling. Most studies suggest that the interphase is most

Table 1. Properties of the materials used in the study

Property	PH	PM	PL	EH	EL	Fibre
E (GPa)	3.22	2.56	1.39	5.67	1.39	76
Poisson's ratio	0.36	0.36	0.36	0.31	0.36	0.25
Yield strength (MPa)	59.9	40.8	18.5	-	-	-
Yield strain (%)	3.62	2.88	2.39	-	-	-
Draw strength (MPa)	49.5	31.7	17	-	-	-
Density (g cm^{-3})	1.3	1.3	1.3	1.27	1.3	2.5

important for provision of long-term and durability properties but since it is the inherent medium for stress transfer then it will also have a considerable bearing on short term/instantaneous properties.

This study examines a variety of model interphase systems, with a wide variation of material properties. These materials are arranged to form the matrix and a discrete interphase system within a FEA model. Examination of the results will allow an understanding of the effect of an interphase on the strain transfer characteristics of composites.

2 EXPERIMENTAL

2.1 *Materials*

To analyse the effect of differing interphase properties on the strain transfer into the fibre five model materials were defined, using properties measured from real polymers used in composite production which have been investigated previously [8,9]. The materials are denoted by two letters, the first indicating the material nature as elastic (E) or plastic (P), and the second describing the nature of the modulus and yield strength as high (H), medium (M) or low (L). These materials were

  i) EH, an elastic material with high modulus.
  ii) EL, an elastic material with low modulus.
  iii) PH, a ductile material with high modulus and yield strength.
  iv) PM, a ductile material with intermediate modulus and yield strength.
  v) PL, a ductile material with low modulus and yield strength.

The reinforcement used was modelled as glass fibre.
Properties of the different materials are listed in Table 1. The model systems were labelled as matrix-interphase respectively (e.g. XX-YY denotes a system consisting of matrix material XX and interphase material YY).

The properties of the materials were incorporated into the FEA model by digitisation of the true stress/strain curves. The 3D stress state in the model was calculated from the 2D stress/ strain curve by the FEA solver using a Von-Mises criteria.

2.2 *Finite Element Model*

A 2D axisymmetric finite element model was produced

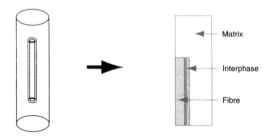

Figure 1. 3D and 2D representations of concentric cylinder interphase model.

to investigate the effect of stress transfer in a short fibre composite. The model also has relevance to fibre fragments in the single fibre fragmentation test. 2D representations of a 3D problem have been shown to give results equivalent to a full 3D solution by previous studies [6] and so provides a valid modelling method.

The 2D representation was based on a concentric cylinder approach, having 3 cylinders, one each for fibre, interphase and matrix respectively. The relevant 2D and 3D models are shown in Figure 1.

The model was produced and meshed using FEM-GEN software, with second order elements of type CAX 8 (8 noded biquadratic 2D strain elements) in order to reach an accurate solution. The model comprised 2048 elements and 6337 nodes. The FEA mesh was graduated to concentrate computational power in the most important area, the fibre end. Away from this the mesh gradually coarsened as fewer stress events are present, requiring less computational effort.

The model geometry was chosen to represent a short fibre in resin. The width of matrix used was selected so that the strain at the edge furthest from the fibre was only 1% of the maximum stress achieved anywhere in the model, hence edge effects are unimportant [9].

The model was solved by ABAQUS [7] using a non-linear static analysis, where each loading increment was iterated for equilibrium using the full Newton method. The results were investigated and documented using ABAQUS/POST [7]. It was constrained on side ABC in the radial direction and side AE in the fibre axis direction. Side CD was loaded axially in 0.5% incremental steps of applied strain upto a total of 2.5% applied strain, so progress of the response could be monitored at defined strain levels.

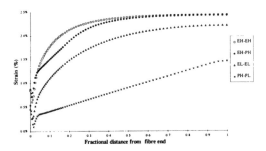

Figure 2. Graph showing typical strain profiles in the fibre centre, for four matrix-interphase systems at 2.5% applied strain, showing the differences when only elastic deformation can occur (EH-EH and EL-EL), and when yielding occurs in both a high yield strain interphase system (EH-PH) and low yield strain interphase system (PH-PL).

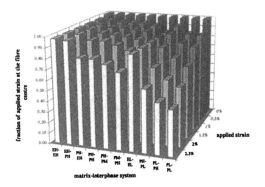

Figure 3. The proportion of the applied strain attained in the fibre centre, or Strain Transfer Efficiency (STE) for each matrix-interphase combination, at each applied strain.

3   RESULTS

Typical strain profiles from within the fibre centre at 2.5% applied strain, for four of the systems used in this study, are shown in Figure 2. All the profiles show an increase in strain from a minimum near the fibre end, to a maximum at the fibre mid-point, although the minimum occurs slightly away from the fibre end as stress transfer may occur across this region. Between the minimum and maximum, strain is developed in different manners depending on the materials present in the interphase and matrix.

The two systems which are purely elastic (EH-EH and EL-EL) show a typical exponential development of strain in the fibre-centre, to a plateau. Near the fibre mid-point, the strain is very close to the applied strain. The deviation from the applied strain is greater in the EL-EL system, because the modulus of the matrix is lower. It can also be seen that the plateau region is shorter. The EH-PH system also exhibits strain at a magnitude close to that applied. However, near the fibre-end, a region of constant gradient can be seen, causing the strain to increase at a reduced rate leading to a shorter plateau. This is a consequence of yielding in one of the constituents, (in this case the PH interphase) leading to a region of constant shear stress. In the case of the PH-PL system, it can be seen that the region of constant gradient is much more extensive than in the EH-PH case. This is because the yielding occurs over a larger fraction of the fibre length, within the lower yield strength PL interphase. This can be seen to have led to a greatly reduced plateau length. The maximum strain in the fibre centre is also much lower than in the EL-EL system.

In order to compare these curves, two values which characterise their overall shape are employed. The first of these is defined as the proportion of the applied strain attained in the fibre-centre and is termed the strain transfer efficiency (STE). This gives an indication of the

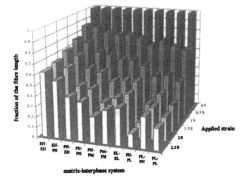

Figure 4. The proportion of the fibre length at or above 95% of the maximum attained strain, or Strain Transfer Length (STL), for each matrix-interphase combination at each applied strain.

efficiency with which the strain in the matrix is transferred through the interphase, into the fibre. The second is a measure of the length of the plateau region leading up to the fibre midpoint. This is defined as the proportion of the fibre length at a strain greater than 95% of the maximum and is termed the strain transfer length (STL). The STL-value gives an indication of the effectiveness of strain development and the load carrying ability of the system. The STE and STL values for each of the matrix-interphase combinations used in this study are plotted in Figures 3 and 4 to allow graphical comparison of the data. The area under the strain profiles (the total strain transferred) was also measured for each system at each applied strain. This data was normalised by dividing the measured area by the theoretical maximum area (applied strain x fibre length), to attain a normalised Strain Transfer Function which is displayed in Figure 5. This data provides an indication of the load carrying

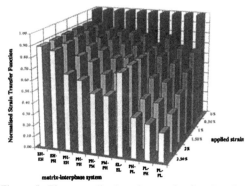

Figure 5. The normalised strain transfer function for each matrix-interphase combination at each applied strain.

capability of the composites, complementing the STE and STL values.

4 DISCUSSION

From Figures 3 and 4 it can be seen that prior to yield in the samples (i.e. 0.5% applied strain), the STE and STL are related to the matrix properties and are largely unaffected by the interphase characteristics. Yielding reduces both the STE and the STL, which can be seen when the EL-EL and PL-PL systems are compared. The elastic properties of these materials are the same but the PL material is capable of yielding. Yielding in these systems is characterised by a fall in the STE value as the applied strain increases. The purely elastic samples show no decline in STE with increasing strain, whereas all of the other samples show some tendency to decrease.

The data reveals some other factors that affect the STE and STL values. It would appear that the STE value depends more on matrix rather than interphase properties. This can be seen when all the systems with the PH matrix are compared. The PH-EH, PH-PH and PH-PM systems all have reasonably similar STE values and the PH-PL system is not greatly reduced despite the large variation in interphase characteristic, i.e. if the interphase stiffness and yield strength are lowered, the STE is not significantly reduced as it is matrix dominated.

The opposite relationship is found to exist for the STL, with system properties found to be more interphase dependent. This can be seen by comparing the PH-PH, PM-PH and PL-PH systems, which are very similar in STL value despite the dramatic change in matrix properties, i.e. if the matrix stiffness and yield strength are lowered the STL is not significantly altered. It is also observed that the PM-PH system has a higher value of STL than the PH-PM system despite the stiffer matrix in the second case.

From the above observations it can be stated that a stiff matrix gives a high STE, whilst a stiff interphase gives a high STL. A stiff matrix is beneficial in maintaining a high STL but the effect is less than that of the interphase. The same is true of a stiff interphase on the STE value. The final observation is that while yielding does adversely affect the strain transfer characteristics, the overall effect is not necessarily significant compared to an elastic situation (see the EH-PH and EH-EH systems). However, two systems can have different STE/STL characteristics and still ultimately transfer the same amount of strain into the fibre. Figure 5 shows that the PH-PM and PM-PH systems transfer the same quantity of strain into the fibre but it is done in different ways, as they have different STE and STL profiles (see Figures 3 & 4). By examining this data, it would be possible to design two composites that would have the same reinforcement efficiency, but would exhibit different mechanical responses. An example of this is a comparison of the EH-EH and EH-PH systems (Figure 5). These show similar strain transfer functions, and therefore reinforcement efficiency, but the interphase in the latter case is able to yield, which would shield the fibre from crack propagation in the matrix or conversely, shield the matrix from a fibre fracture. The totally brittle EH-EH system would fail catastrophically as there would be no barrier to crack propagation and no capacity for stress relief.

The implications of these observations to composite materials are important. Considering the fragmentation test geometry, the central portion of the fibre will experience less strain if yielding occurs. Additionally the length over which the strain is maximum will be reduced, and therefore yielding will cause an increase in the final fragment length. Thus, in analyses where yielding of the matrix is not considered, a large error in the interfacial shear strength would be introduced.

In the case of real composites, the implications are equally profound. This study has shown that where a soft interphase is present, the STE and STL value is reduced. Therefore, a yielding interphase grossly reduces the strain transfer efficiency and thus, the reinforcing efficiency of the fibres. However, yielding at the interface has been shown to relieve strain in the fibre and dissipate energy. Thus, yielding may help to increase the overall strength/toughness of a composite.

5 CONCLUSIONS

The proportion of the applied strain attained in the fibre centre, or Strain Transfer Efficiency (STE) and the proportion of the fibre length at greater than 95% of the maximum strain, or Strain Transfer Length (STL) are shown to be a simple means of characterising strain transfer processes in composites, allowing the comparison of different matrix/interphase systems. Using this approach, the effect of yielding on the strain transfer efficiency has been analysed for various model systems containing differing matrix-interphase pairs. The properties used in the model were those determined experimentally from matrix resins employed for composite materials (see Table 1). Overall, greater yielding was found to give lower STE and STL values. Therefore, yield clearly has an adverse effect on the strain transfer processes within composites, but may increase toughness. The STE was shown to be matrix property influenced whilst the STL was interphase property influenced.

6 ACKNOWLEDGEMENTS

The authors would like to thank EPSRC and The Advanced Composite Group for funding, and the European Community Interphase Project.

REFERENCES

Jones F.R., Key Engineering Materials, 116/117, 41, 1996.

Drzal L.T., Mat. Sci. & Eng., A126, 289, 1990.

Swain R., Reifsneider K.L., Juyaraman K., & El-Zein M., Thermoplastic Composites, 3, 13, 1990.

Reifsnider K.L., Composites, 25(7), 461, 1994.

Hsi Chin Tsai, Arocho A.M., & Gause L.W., Mat. Sci. & Eng., A126, 295, 1990.

Weissenbek E., & Rammerstorfer F.G., Acta Mettall. Mater, 41(10), 2833, 1993.

Hibbit, Karlsson, & Sorenson; Rhode Island, USA.

Chen F., Tripathi D., & Jones F.R., Composites Part A, 27A, 505, 1996.

Mader E., Jacobash H.J., Grundle K., & Geitzelt T., Composites Part A, 27 A, 907, 1996.

Experimental Techniques and Design In Composite Materials 4, Found (Ed.)
© *2002 Swets & Zeitlinger, Lisse, ISBN 90 5809 370 0*

Author index